Fixed Income
Mathematics

Fixed Income Mathematics

Robert Zipf

ACADEMIC PRESS
An imprint of Elsevier Science

Amsterdam   Boston   London   New York   Oxford   Paris
San Diego   San Francisco   Singapore   Sydney   Tokyo

Academic Press
An imprint of Elsevier Science
525 B Street, Suite 1900, San Diego, California 92101-4495, USA
http://www.academicpress.com

Academic Press
84 Theobald's Road, London WC1X 8RR, UK
http://www.academicpress.com

Library of Congress Catalog Card Number: 2003102997

International Standard Book Number: 0-12-781721-2

PRINTED IN THE UNITED STATES OF AMERICA
03   04   05   06   07   8   7   6   5   4   3   2   1

CONTENTS

20 Variable and Uncertain Cash Flows

21 Mortgage-Backed Securities

CHAPTER 1

Introduction—Who this Book is for and What it Hopes to Accomplish

HISTORICAL BACKGROUND—THE BIG CHANGE IN INVESTMENT, LOAN, AND MONEY MANAGEMENT

Any person who has been active in finance and investments during the last 45 years is surely aware of the enormous changes that have occurred in the area of personal financial management during that time. A truly immense range of opportunities for both borrowing and investment has opened up for most people in this country. But these opportunities have also presented financial management problems. This book tries to give the reader the mathematical tools to solve these problems.

In the 1950s, investments for most people consisted of deposits in savings accounts and permanent plan life insurance. A few adventurous investors bought common stocks, but not many. Stock trading costs were much higher then than now. Keith Funston, then President of the New York Stock Exchange, actively encouraged individual ownership of stocks, and Merrill

Lynch introduced the monthly investment ownership of stocks, and in January 1954, Merrill Lynch introduced the monthly investment plan. The monthly investment plan made it relatively easy for individual investors to buy stocks by making a monthly investment of as little as $40. However, some on Wall Street thought that individuals should not buy stocks at all, but rather content themselves with the modest returns, but complete safety, of savings accounts and permanent plan life insurance.

Borrowing was similarly simple. Prospective home buyers visited their local bank, and, if their application was approved, received their mortgage from the bank. The income of working wives was not considered in the family income for this purpose; presumably, the wife would quit work, have babies, and become a full time housewife. The bank continued to own the mortgage, and received the payments from the borrower. Second mortgages, now called home equity loans, were rarely taken out and were financed by a separate set of companies, frequently run by individuals of somewhat dubious reputation, if not of dubious character. The very existence of a second mortgage was taken as evidence of financial mismanagement at least, and possibly of worse.

If the borrower wanted a consumer loan—to purchase an automobile or refrigerator, for example—he went to one of the local consumer finance companies and received his loan, promising to make a series of monthly payments. Sometimes, as now, the seller of the product offered to finance the purchase. The borrower usually had no idea of the true cost of the funds he borrowed; indeed, the author knew one principal of a locally well known consumer finance company who himself had no such knowledge. (The author bought him a book of compound interest tables, together with some problems to solve. The principal spent considerable time working out the problems, much to the amusement of his father, a major owner of the firm. His father thought such knowledge was not relevant to the main business of the firm. He could have been right.) The borrower made these payments directly to the lending company. Choices were few compared to the many available today.

Pensions were similarly simple. If an employee worked for many years for a single company, the company would grant him a pension, if it had a pension plan. The pension's size depended on the employee's salary, length of company service, and age at retirement; this method is called a defined benefit plan. Some pensions never vested; if the employee left before retirement age, he did not receive a pension. The very good plans had a 10-year vesting provision; after 10 years of service, the employee was guaranteed a pension at retirement age.

All of these policies have changed. Common stock purchase is widely promoted, and stocks are easily traded, at much lower commission rates than those of 50 years ago. Mortgages are now packaged, and few banks keep them in their own portfolios. Some mortgages are even offered on the Internet.

Similarly, mortgages are easily refinanced when interest rates fall to levels that would make refinancing worthwhile. Home equity loans are heavily promoted and easily obtained by those with suitable credit. As a result of government action, interest costs are stated, in annual percentage rates, for most loans. Companies are now moving away from defined benefit pension plans toward defined contribution plans. In a defined contribution plan, the employee makes contributions into one of several types of investment accounts offered. Employer contributions are usual, up to a limit. Management of the employee's retirement funds is now the employee's own responsibility. (Your pension is your problem.) The vesting period is now 5 years for most plans. Note also that during the presidential campaign of 2000, private investment plans were seriously proposed as part of Social Security. Individual retirement arrangements (IRAs), Keogh, and similar plans, as well as other retirement plans, have increased the choices available and have similarly increased the problems.

These new choices have offered most of us a wide range of financial management possibilities and responsibilities. But the educational system has not kept pace with the requirements in this area. The mathematical tools available to solve the problems have been relegated to bond analysts, actuaries, and project analysts and managers. Perhaps fixed-income mathematics should join driver's education and typing, now called "keyboarding," as a normal public school educational offering.

The views of investors have also changed somewhat during this period. Investors used to look at a bond as an income-producing investment. The size of the income was somewhat related to the amount invested, and it was either spent for living purposes or reinvested, mostly in other bond investments. When the bond matured, the proceeds were reinvested in another bond. Today, financial professionals frequently look at a fixed-income investment as a set of independent single-payment cash flows, each with its own individual value.

But the first view is still very much with us. A good friend of mine, head of municipal research and recently retired from a major Wall Street wire house, firmly believes that you buy a bond to provide a cash flow. You use the cash flow to meet your spending needs. When your bond matures, you buy another bond. Most investors probably have this view.

WHAT THIS BOOK HOPES TO ACCOMPLISH

With this book, we hope to give the reader the mathematical tools necessary to analyze and solve most problems in the mathematics of finance that one is likely to encounter. It will also give references for further study if the reader is interested and has the required mathematical knowledge. These references will most likely be in the fields of derivatives and options. A full understanding of the mathematics of these fields requires knowledge of partial differential equations, probability and statistics, and stochastic differential equations, areas of mathematics far beyond the knowledge of most investors.

WHAT SORT OF PROBLEMS MIGHT THIS BOOK HELP YOU TO SOLVE?

Here are some examples of the sort of problem that you might easily face and that are covered in this book.

1. You are borrowing on a mortgage to buy your first home. The bank has offered you two choices for a 30-year mortgage: you can borrow the full amount at 7%, or you can borrow at 6.75% with two points (a 2% reduction in the cash you actually receive from the face amount of the mortgage). You plan to stay in the house for many years. Which choice should you pick and why?
2. You now plan to live in the house only five years before you move. Does this change your response to Problem 1 and why?
3. Your broker has offered you two investments, a 6-month Treasury bill at a discount yield of 3.92% and a Treasury note due in 6 months at a bond yield of 4%. Which should you pick and why?
4. Your broker has told you that your fixed-income portfolio has a duration of "only 3 years." What does this mean, why should you care, and what, if anything, should you do about it?
5. You manage the plant for a major widget manufacturer. The company's chief financial officer has said that he expects each new project to yield a 10% annual return. Your assistant has proposed a project to improve production. The project will cost $50,000, has a 90% chance of yielding $10,000 per year for 10 years, and has a 10% chance of a total loss. Should you do the project and why?

WHO THIS BOOK IS MEANT TO ADDRESS

This is an introductory book meant for anyone who wishes to analyze problems, of any sort, that involve the present value of a future flow of funds or any other goods or services that can be evaluated. It should help the reader to make decisions and plan future courses of action. Investment will usually account for a large segment of these problems, but by no means the only one. The book discusses some nonfinancial factors that might influence the final decision. The book contains a variety of examples from many different fields to give the reader an idea of the truly wide range of problems that can be analyzed using the concepts presented in this book.

Those who wish to choose intelligently among the various ways of borrowing or among the various investment opportunities available to them will benefit from the methods presented in this book. Likewise, people who wish to choose intelligently among various business management or investment choices well find much to offer here. The book should give these readers some ideas on how to solve their problems, as well as some idea of other, non-mathematical factors involved.

This book is also meant for those with a strong mathematical knowledge, but with no knowledge of financial mathematics, who wish to learn to apply higher mathematics in the financial field. These might include new Ph.D.s in math who have no real knowledge of financial mathematics. These readers will not find the financial math in this book terribly difficult or profound, but it still must be studied and learned if the reader wishes to apply it; for them, the book might be an evening's read. They can then go on to their main field of interest, which could easily require a much higher level of mathematical knowledge than this book demands. A list of references is provided, but most practitioners will already have a list of books, possibly provided by their employer.

THE MATHEMATICAL KNOWLEDGE REQUIRED FOR THIS BOOK

This book assumes that the reader has a reasonable facility with the mathematics studied in high school by most people. The book requires some familiarity with mathematical equations, how to manipulate them, and how to understand and use them. The ability to cast problems in mathematical terms is almost a requirement as well. Knowledge of calculus is helpful, but not required. However, this knowledge would provide better insight into financial mathematics. As the world becomes more technical, and mathematics and other scientific knowledge becomes more widely used, knowledge of mathematics becomes more desirable for many purposes. If you don't know calculus, perhaps you should learn it.

Some parts of the book require a mathematical understanding beyond the scope of the book for rigorous development. I have tried to offer enough to allow readers to understand what is being done without burdening them with mathematics beyond their interests. These areas include, for example, numerical analysis and the real and complex number systems. References are provided if the student wishes to do further study. Occasionally, several different approaches might be used to reach the same solution. For example, I use three different approaches to the subject of continuous compound interest functions.

Reader understanding of the principles is absolutely important. The text is designed to further that understanding. For example, I use the bisection method for numerical approximation solutions to equations. Mathematician readers have pointed out, correctly, that much faster numerical analysis algorithms exist. That is true, but the bisection method is easy to understand and to use. You can easily explain it to your boss and to customers. Other, possibly better, methods properly belong in a course on numerical analysis.

THE ROLE OF EXAMPLES AND PROBLEMS IN THIS BOOK

This book is designed to be used as a text, so plenty of examples and problems are provided. These range from simple calculation, using the equations provided in the applicable lesson, to problems that require not only math, but consideration and possible class discussion of the techniques needed to solve the problem. These problems might not even have an answer, but simply a statement of how the problem might be solved. The author believes that the problems themselves will be a learning experience.

Most chapters have three kinds of problems. One set, "Topics for Class Discussion," offers subjects, frequently of current interest, for the class to discuss. Discussion might include methods of solving the problem and approaches to take, as well as a possible actual solution. Some of these topics include discussions of public policy on such proposals as building a new Tappan Zee Bridge and revision of the Social Security program. A second set will simply be problems to solve, with the student obtaining the (or a) correct answer. A third set is designed for computer usage.

This book is based on a fixed-income mathematics seminar that the author has given at the New York Institute of Finance, now FT Knowledge Financial Learning, in New York City, since 1993. The format and the material covered seemed to meet a need, at least in the Wall Street area of New York City.

I hope that students will realize the truly wide range of activities where fixed-income mathematical analysis can help them to make decisions. This can especially happen in cases in public policy. A wide range of alternatives may be available for various public programs. Public analysis of these options could result in better public decision making, higher overall public awareness of the various options, and a better use of public funds in achieving overall public objectives. I hope that this knowledge will lead to better citizenship and better public policy decisions. This hope has been a major driving force in preparation of this book.

CHAPTER 2

Interest, Its Calculation, and Return on Investment

This chapter introduces the concept of interest as payment for the use of capital, first in money form, with examples. It then presents the equation for calculating simple interest. The chapter then expands the interest concept to the notion of return on investment, with a wide variety of examples. The examples are particularly meant to make you aware of the various places where you can use financial analysis. You should especially know the conventions in the United States of interest and yield calculations on bonds and mortgages. When you finish this chapter, you should understand the concepts of interest and return on investment and have some knowledge of the many places where they can be used.

A GENERAL INTRODUCTION TO INTEREST

Interest is payment for the use of capital, usually in the form of money. It can also be thought of as rent for money borrowed for a period, which must be returned at the end of the period. Interest is stated as a percentage per time period, frequently in the form of an annual rate.

Most loans have a date on which the amount of the loan becomes due and payable to the lender. This is called the maturity date, and the loan is said to mature on that date. Some loans, such as most home mortgages, have planned periodic (usually monthly) repayments of principal before the loan matures. For most mortgages, these scheduled principal repayments are part of the monthly payment, which also includes interest. The maturity date of these loans, usually called the final maturity, is the date when the final mortgage payment is due.

EXAMPLE 2.1.   Your neighbor borrows $100 from you to help pay his taxes. He promises to repay you in three months and to pay you interest at an annual 6% rate.

Here the interest rate is 6% per year. The loan comes due in 3 months; we say the loan matures in 3 months, or has a maturity date 3 months from now. The principal amount is $100.   ■

EXAMPLE 2.2.   Shortly before May 15, 1995, the United States Treasury sold some 10-year notes. They pay interest at an annual rate of $6\frac{1}{2}\%$ (6.5%) and come due (mature) on May 15, 2005. Here the interest rate is 6.5% per year, paid in two semiannual installments, and the maturity date is May 15, 2005. This stated interest rate of the notes is also called the coupon rate, and the payments are called coupon payments.   ■

EXAMPLE 2.3.   Many readers may own United States savings bonds. These bonds are sold below their face amount (called par) and mature at par years later. The interest is accumulated and is paid when the owner redeems the bond, at maturity or whenever the owner chooses to redeem them (provided the owner has held them for the required period).   ■

HOW TO COMPUTE INTEREST

The traditional equation for interest computation is

$$\text{Interest} = \text{Principal} \times \text{Rate} \times \text{Time}$$
$$I = PRT$$

This is called "simple interest." Compound interest is the usual basis for extended calculation, and it is discussed in the next chapter.

EXAMPLE 2.4.   In the case of Example 2.1,

$P = 100$, the amount you lent to your neighbor
$R = .06$, the 6% per year rate you are charging
$T = .25$, one-quarter year, or 3 months

If your neighbor borrowed for 1 year, the annual interest he owes you would be

$$I = PRT = 100 \times .06 \times 1 = 6.00$$

However, he is borrowing for only 3 months, or one-quarter year, so the interest he owes you will be

$$I = PRT = 100 \times .06 \times .25$$
$$= 1.50$$

EXAMPLE 2.5.   In Example 2.2,

P = The par amount of notes you own
$R = .065$
T = The length of time you own the notes

The interest cannot be calculated until P and T are known. Also, Treasuries have their own special conventions for computing accrued interest. These will be discussed later in the book.

However, suppose you own $10,000 par amount and hold them for 1 year. Then

$$P = 10,000$$
$$R = .065$$
$$T = 1$$

I equals 650. This will be paid in two semiannual installments of $325 each.

Note that we have used only numbers in these calculations. Many people prefer to think in terms of dollars and cents, and actual money transactions. Others prefer to think solely in terms of the numbers. Both will produce the same results. Use the method most convenient to you.   ■

NOTATION

Here is the notation we will use throughout this book. Other financial mathematics books may have different conventions.

I The actual amount of interest. This will, or can, usually be expressed in money terms.

   i The rate of interest, expressed as a decimal.
   y Yield, the term used in most equations in the bond business. Bond
        yield is equivalent to i above.
   P Price. Bond prices are expressed as a percentage of par, not (usually) as
        dollars per bond.
   S The actual amount of money, usually principal, in the equation. This,
        along with I, will or can usually be expressed in money terms.
   t Time

Using the new notation, our equation

$$I = PRT$$

becomes

$$I = S \times i \times t$$

PERCENTAGE RATE AND TIME PERIOD

The interest rate is often expressed as a percentage rather than as a decimal.
In this case, to do the calculation, divide the percentage rate by 100 (convert
it to a decimal form).

   EXAMPLE 2.6.   In Example 2.1, we converted 6% to .06, as follows:

$$6/100 = .06$$

The interest rate description must contain both a rate and the time period
to which that rate applies. The problem does not make sense, and cannot be
solved, without both the rate and the time period. The time period is not
necessarily the same as the period the loan, or other flow of funds, will be
outstanding.   ■

   EXAMPLE 2.7.   In Example 2.1, the rate is 6% and the time period is 1
year, even though the loan will mature in just 3 months.   ■

   EXAMPLE 2.8.   You are offered an investment with a fantastic 15% return.
After a few questions, you find that this 15% return is paid as a lump sum after
5 years. It does not look like such a great return, and you decline to invest.
                                                                          ■

   EXAMPLE 2.9.   You are offered a loan at a low 3% interest rate. You do
some investigating and find that the rate is 3% per month. You decide that the
rate does not look so low, and you elect not to take the loan.
   These rates and time periods are not always explicitly stated. Sometimes a
trade custom changes the actual rate and time period. Here are two important

examples from the United States. You should know both of these trade customs.   ■

    EXAMPLE 2.10.   In the United States, bond yields and bond coupon rates are stated as annual rates, but the annual rates are compounded semiannually. In April 2002, the long-term Treasury bond yielded about 5.8%. This does not mean that it yields 5.8% per year. It means that it yields 2.9% per half year, and we say that the bond has a nominal annual yield of 5.8%, compounded semiannually. This will give somewhat different answers than the plain annual rates. Consider the Treasury note in Example 2.2. If you paid 100 (par) for the note on an interest payment date, your broker would tell you that you bought it at a 6.5% annual yield. Your broker actually means that you bought it at a 3.25% semiannual yield. We say that the note yielded you a nominal annual rate of 6.5% per year, compounded semiannually. Almost all bonds traded in the United States of America have their yields stated as nominal annual rates compounded semiannually. In other countries, other trade customs apply. For example, in Germany, the rates quoted are true annual rates.   ■

    EXAMPLE 2.11.   Many readers will have a mortgage on their home. Most mortgages issued in the United States have a fixed interest rate. The interest rate on the mortgage is usually stated as an annual rate. However, almost all home mortgages in the United States have monthly payments. This means that the actual rate paid is not the usually stated annual rate, but the annual rate, divided by 12, compounded monthly. For example, if the stated rate is 8% annually, the actual rate is $(8\%)/12 = (2/3)\%$ per month. The nominal annual rate of 8% is compounded monthly.

    Many loan advertisements show both the nominal and actual percentage rates. For example, the real estate section of the Sunday edition of a leading East Coast daily shows both the nominal and annual percentage rate (APR) of the loans advertised. These APR quotes for mortgages include both the effects of compounding and the deduction of points.   ■

RETURN ON INVESTMENT

The discussion and examples shown so far have all involved an actual payment of money and a stated return of money. The money return is due on specific dates, and these terms are all stated in the loan contract at the beginning of the loan.

    This can be expanded to a general idea of an investment of some sort and a return of some sort. The investment need not be entirely money, and the return need not be entirely money either. However, the same mathematics applies in these cases.

What sort of investment activity can be covered? Examples include investments that produce savings, but no cash flow; investments of time, labor, or property, with or without cash; investments that produce a noncash return; and similar investments. Here are some examples.

EXAMPLE 2.12.   In ancient Mesopotamia, farmers would borrow grain, frequently from the temple, and repay the loan with a larger amount of grain at harvest time. The code of Hammurabi (ca. 1800 B.C.) regulated the amounts that the lender could charge for these loans. In this case, the investment and return were both in the form of grain.   ■

EXAMPLE 2.13.   Your neighbor plans to build a vacation cottage on an upstate lake. He invites you to help him build the cottage, and in return he will let you use it, rent free, for 2 weeks each summer. Here your investment is your labor in helping build the cottage, and your return is 2 weeks of free use each summer.   ■

EXAMPLE 2.14.   During World War II, meat was rationed. Two families made a deal. One family bought a piglet, and the other family raised the pig. When the time came to slaughter the pig, the pig was taken to a local slaughter house, and each family got one-half of a full-grown pig, fully slaughtered and cut up into pieces suitable for cooking. In this case, the investment of the family that bought the piglet was

The cost of a piglet
One-half the cost of slaughtering the full-grown pig

The investment of the family that raised the pig was

The cost of raising the pig (the cost of the grain feed)
One-half the cost of slaughtering the full-grown pig

The return on the investment for each was one-half a full-grown pig, fully slaughtered and cut up. (Note: This really happened.)   ■

EXAMPLE 2.15.   At one time, New York State had a program that allowed veterans to use certain funds from military bonuses, disability payments, and other sources, called "eligible funds," to pay down loans used to purchase their homes. At the same time, this reduced the assessed value of the house for certain real estate tax purposes by the amount of the applied eligible funds, up to $5,000. The proceeds from surrendering a GI insurance policy were considered eligible funds. A veteran cashed in (surrendered) this type of GI insurance policy when its cash surrender value, together with the accumulated dividends, reached $5,000, and then paid down his mortgage by this amount. This reduced both the amount of interest paid on the mortgage (because the mortgage was smaller) and the amount of real estate taxes due on the house (because the assessment was reduced for tax purposes). The investment and return on investment were as follows:

> Investment: $5,000 cash surrender value of the policy and dividends, used to pay down the mortgage by this amount.
> Return: $300 annual interest savings on the mortgage, which carried a 6% interest rate.
> Annual savings on real estate taxes: These savings change each year, and usually increase. For the year 2000, the veteran saved about $1,270, or about 25% of the $5,000 investment. ■

EXAMPLE 2.16. You are the manager of a manufacturing plant. You can buy a new machine for your production line, which will reduce production costs. The machine costs $25,000 and will save you $10,000 yearly in costs. In this case, the investment is $25,000 cash, and the return is a saving of $10,000 per year. Some of this saving could be in avoidance of cash expense, but other savings might be in avoidance of other charges, such as depreciation.

When you determine the investment and the return, make sure you include all the related costs and returns of each. For example, in Example 2.13 (the lakeside cottage), additional investment might come from several sources. If you drive yourself and your neighbor to the site, you could add travel costs to the investment. If you treated him to lunch, this might be considered additional investment. If he treats you to lunch, this could be considered an additional return on your investment. In Example 2.14 (the pig), the costs of transporting the feed grain from the grain dealer to the family grain storage area might be additional investment. In Example 2.15 (the GI insurance), additional investment might be the give-up of the cost savings on low-cost GI insurance. If the $5,000 could be invested in Treasury obligations at an 8% yield, rather than used to pay down the 6% mortgage, the extra 2% per year could be considered part of the investment or, alternatively, used to reduce the total return. In Example 2.16, if the machine had shipping costs of $1,000 and installation costs of $500, then the total investment would be $26,500, as follows:

Machine cost	$25,000
Shipping	1,000
Installation	500
Total	$26,500

Sometimes some of the costs are so small that their inclusion will not change the final decision. The costs of postage to mail a check to pay down a loan would not usually be considered in the analysis. In Example 2.15 (the GI insurance), the annual 2% costs were much smaller than the real estate tax savings. However, you should always try to include all relevant costs and returns before making a final investment decision.

Note that the word "return" can be somewhat ambiguous. It can mean everything you get back from your investment, including the value of the original investment. However, it could also mean simply the profit, or earnings, from the investment. For example, if you buy a bond, receive the interest until the bond matures, and get your money back when it matures, your return and your interest earnings are the same. However, in the case of the pig, the return of one-half a full-grown pig included the original investment of the piglet or grain feed for the pig. The context will usually show the correct meaning of the word.   ■

ANALYSIS OF INVESTMENTS OR RETURNS WITHOUT EXPLICIT MONEY VALUES, AND INTANGIBLE INVESTMENTS AND RETURNS

Sometimes you make an investment or receive a return that is not expressed in money but is important in the decision process. Try to assign reasonable values to these items.

In Example 2.13 (the lakeside cottage), you might figure the value of your labor and the value of the cottage rental. Suppose you earn $25,000 per year, or $500 per week, equivalent to $100 per day. Suppose also that you spend your two-week vacation and 10 two-day weekends helping your neighbor. You might have worked two weeks ($1,000) and 20 additional days (another $2,000), for a $3,000 total. If similar cottages on similar lakes rent for $200 per week, then the value of your free use is $400. You are receiving $400 each year in value on an investment of $3,000 in value.

Often these intangibles will have some element of judgment. For example, in the preceding case, should you use before-tax or after-tax income? Should you include the Social Security and Medicare deduction? Sometimes these factors can be important.

In many cases, the intangible return may be quite important, but not measurable, or at least not easily measurable. For example, many people find that paying off their home mortgage results in a level of peace of mind and general contentment that is important to them, but cannot easily be evaluated.

In Example 2.14 (the pig), the intangible benefits would include having plenty of meat to offer guests without worrying about running out, especially with meat rationed during the war. What is it worth to avoid this kind of worry?

In Example 2.13 (the lakeside cottage), intangible benefits might include avoiding the bother of making rental arrangements for a summer vacation. You have already made the arrangements with your neighbor. Note that avoidance of work or bother may often be a return on an investment. Here are two more examples of intangible benefits as a return on investment.

EXAMPLE 2.17.   *Investment in tax-exempt municipal bonds.* Some studies by the Internal Revenue Service have found that many individuals own tax-exempt municipal bonds but do not benefit from the tax-exempt feature of these bonds. They would earn more money after taxes if they sold the municipal bonds, bought taxable bonds, such as Treasuries, and paid the income tax due. Why might someone own municipals in such a case? Perhaps the feeling of having some income that the federal government cannot get its sticky fingers on compensates for the lower actual cash spending income. Or perhaps one might give a spouse some municipals. The spouse would then receive income, which he or she could spend on lunches with friends, Christmas presents for the children, or jewelry, without worrying about the tax due on this income.
■

EXAMPLE 2.18.   *Bank service charges.* Many banks will allow a depositor to avoid bank service charges on his or her checking account if the depositor maintains a certain minimum balance. For example, one large East Coast bank will waive all service charges if the checking account has a minimum balance of $6,000 or if the customer has a total balance of at least $25,000 in all his or her accounts. The customer who maintains the minimum balance has a tangible return of the amount of the waived service charge. The customer also has the intangible value of avoiding a pesky, relatively small, charge on his or her account.

Sometimes you can figure the implied value of the intangible benefits, at least to the person paying for them. Usually this requires an equivalent investment that yields a cash return.   ■

EXAMPLE 2.19.   *Bank service charges, revisited.* Suppose Susan maintains $6,000 in her checking account solely to avoid bank service charges, plus an additional amount to ensure that her balance never falls below $6,000. She figures that she avoids $15 in monthly service charges, or $180 per year. In April 2002, Susan could invest the $6,000 in 1-year Treasury obligations, yielding 2.5%, or $150 per year. ($I = Sit$, remember? $I = \$6,000\ (S) \times .025$(per year) $\times 1$ (year) $= \$150$.) However, Susan must then pay $180 annually in service charges. If Susan keeps the $6,000 in her account, she is receiving $30 ($180 in service charges less $150 in foregone income) in value each year and avoiding the bother of service charges.

You can improve this analysis by including the effects of federal income tax. Suppose Susan is in the 28% income tax bracket, a common bracket for taxpayers. She would pay $42 in federal income tax on the interest income, leaving $108 available as income. (U.S. Treasury interest is exempt from state income taxes.) This is even more advantageous to Susan, in addition to avoiding the bother of the bank service charges.

If Susan could earn 5% on the $6,000, she would earn $300 annually. After the 28% income tax, she would have $216 left over. She is paying $36 annually to avoid the bank service charges. Is it worth it? It is to Susan. Someone else might rather pay the service charges and have $3 in extra monthly spending money.

When analyzing intangible returns, don't go overboard and overestimate these values. They can dissipate, your situation can change, and there is no money involved anyway. Always be especially conservative in evaluating these returns. You may "feel like a million dollars" when you pay off your mortgage, but that doesn't mean you have an actual million dollars to spend.   ∎

CHAPTER SUMMARY

Interest is payment for the use of capital.
It may also be looked at as return on investment.
It must be expressed as percent per time period.
Both the rate and the time period must be stated for the problem to make sense.
These are sometimes not explicitly stated.
Two important conventions in the United States are

   Nominal annual rates compounded semiannually for bonds
   Nominal annual rates compounded monthly for home mortgages

The equation for simple interest is

$$I = S \times i \times t$$

Make sure your analysis includes the investment costs and returns.
Include intangible costs and returns to the extent possible and reasonable.

TOPICS FOR CLASS DISCUSSION

1. You are off to college in the fall. How would you determine your investment in a college education, and how would you measure the

return? Include whatever intangible costs and benefits you think are relevant. Don't forget to include the earnings income you give up while you are at college.

2. The college has decided to give you student aid: a small scholarship, a large loan, and a job. You must work 15 hours each week, at a college job, for $5.83 per hour. The loan will carry a 6% interest rate, starting after you graduate. How would these changes affect your analysis in Topic 1?

3. At the end of World War II, Congress passed a law, called the GI Bill of Rights, which, among other provisions, gave returning veterans substantial aid toward a college or post-graduate education. From the point of view of the public, what was the investment and what was the return to the public from this provision? Don't forget the public benefits of a better-educated citizenry. How would you quantify this benefit?

4. What was the investment and return from the point of view of the returning veteran who went to school under the GI Bill? Don't forget to consider the point of view of those who might never have even thought about going to college and who had a whole new world opened up for them. How would you quantify this factor?

5. In the case of the pig discussed in the chapter, what if the 9-year-old son of the family fed the pig, along with some chickens and ducks, before he went off to school? Should the value of his labor be included in the analysis? How would you figure the value of the labor if you wanted to include it?

PROBLEMS

1. You have just bought a 30 year 6% Treasury bond at par (face amount, or 100% of face amount). What is the interest rate the Treasury will pay you, both the nominal rate, and the actual rate with its period?

2. You have just borrowed $200,000 to buy a house. The lender is charging you 7.5% interest, with no points (you are receiving the full $200,000). You will make a monthly payment on the mortgage loan. What is the nominal annual rate and the actual rate charged, with its period?

3. You have bought a $1,000 corporate bond with an 8% coupon and have held it for 3 months. How much interest have you earned?

4. You have bought a New York State municipal bond with a $10,000 face amount (par value) and a 4.5% coupon and held it for 2 years. How much interest have you earned on the bonds?

5. You have just bought a condominium in Florida for investment purposes. You paid $80,000 for the condominium, and you have a 2-year lease with a tenant. The tenant will pay all the expenses for the condominium and in addition will pay you $500 each month for rent. What is your investment, and what is your return on this investment?

6. You just realized that you made five air trips to Florida and had four car rentals in order to purchase the condominium in Problem 5. The air fares totaled $2,500, and you paid $700 for the car rentals. You also had to pay a legal fee of $500 when you closed the transaction. Should you revise your investment and return amounts, and, if you decide to revise, what are the revised investment and return amounts?

7. What intangible benefits might you also receive from this investment?

8. A neighbor has offered you an investment that promises a 20% return. What additional information do you need at the very start?

9. Your broker has just bought you a corporate bond at a stated yield of 9% annually. What is your true yield, both the rate and the time period?

10. What is the equation for simple interest?

CHAPTER 3

Compound Interest

This chapter and the next four chapters cover the five standard and most commonly met compound interest functions. A chapter on bond price calculation follows the chapter on annuities certain.

This chapter covers the fundamentals of the compound interest equation. It also covers some mathematical techniques that will be used throughout the book. We develop these techniques together during the chapter so that the compound interest calculations will illuminate the mathematics used. You should understand both the interest equations and the mathematical techniques if you want to get the most out of this chapter. Even if you don't have a strong math background, you should at least have an intuitive understanding of the mathematical concepts involved in this chapter.

When you finish this chapter, you should understand how the compounding of interest works and the equations for interest compounding. You should also understand compounding within a period and the meaning of a nominal annual rate, compounded periodically. Depending on your mathematical skills, you should have at least an intuitive understanding of continuous compounding and when it might be used. You should be able to use compound interest tables to solve problems, and you should be able to solve most problems involving only compound interest calculations. You should understand mathematical models and when they might be used. You should understand the fundamentals of the bisection method of numerical approximation for the solution of many equations. You should be able to use tables to approximate the solutions to compound interest problems.

The equations and mathematical concepts in this unit are the foundation for all our work in this book. It is absolutely essential that you understand the equations and the mathematical thinking behind them before you move on to the rest of the book. You simply will not be able to follow the book if you do not understand these equations or the mathematics.

WHAT IS COMPOUND INTEREST?

Suppose you have $100, and you put it in the bank for 1 year. The bank promises to pay you 8% per year interest. At the end of 1 year, you will have

$$\text{Interest} = I = S \times i \times t = \$100 \times .08 \times 1 = \$8$$
$$\text{Principal still on deposit} = \$100$$
$$\text{Total amount on deposit} = \$108$$

You can also compute the total amount on deposit after 1 year as follows:

$$\begin{aligned}\text{Total amount after 1 year} &= S_1 \\ &= S + I \\ &= S + Sit \qquad\qquad\qquad \text{Equation 3.1} \\ &= S + Si \quad (\text{since } t = 1) \\ &= S(1+i)\end{aligned}$$

In the preceding case, $i = .08$ in decimal form, and $S = 100$, so we have

$$\begin{aligned}S_1 &= 100(1 + .8) \\ &= 100(1.08) \qquad\qquad\qquad \text{Equation 3.2} \\ &= 108\end{aligned}$$

If the total amount at the end of 1 year is left on deposit for the next year, the total time is 2 years, and the equation becomes

$$\begin{aligned}S_2 &= S(1.08)(1.08) \\ &= S(1.08)^2 \\ &= S(1.1664) \\ &= 100(1.1664) \\ &= 116.64\end{aligned}$$

Here is another way to figure the total:
On deposit at the end of year 1: $108.00

During year 2,

$$\$100.00 \text{ earns at } 8\% = 8.00$$
$$\$8.00 \text{ earns at } 8\% = \underline{\;\;.64}$$

Total earnings during year $2 = 8.64$
Total amount on deposit at end of year $2 = 10.8 + 8.64 = 116.64$

This process of the interest paid earning interest on its own is called "compounding," and the result is called "compound interest." The $.64 earned during the second year on the interest paid during the first year is called "interest on interest," and is the result of compounding. If simple interest were used, the result would be only 116.

The $.64 doesn't look like much. Actually, over a period of years, interest on interest can amount to a large sum. In fact, given enough time, there is no limit on how much it can amount to.

If you leave the deposit with the bank for a third year, and the bank still pays 8% per year, the total amount at the end of the third year will be

$$100(1.08)^2 (1.08) = 100(1.08)^3$$

If you leave the deposit with the bank for t years, and the bank continues to pay 8% per year, the total amount will equal

$$100(1.08)^t$$

where t = the number of years you leave the deposit with the bank
If the bank pays i% per year, this equation becomes

$$100(1+i)^t \qquad\qquad \text{Equation 3.3}$$

where

i = annual interest rate, expressed as a decimal, and
t = the number of years (or periods) you left the amount on deposit.

USING COMPOUND INTEREST TABLES

On pages 42 to 49, you will find a set of compound interest tables for various compound interest equations. Few people use these now; computers have almost entirely replaced them. But they are still useful for illustrating how the functions change when one of the variables, such the interest rate or time, is changed. The total of 40 pages shown in this book is quite small compared to the size of compound interest tables published in the 1950s, before computers came along. The author owns one book, published by Financial Publishing Company in Boston, that has well over 1,000 pages. These tables can still be useful for reference or teaching purposes.

The tables show various interest rates along the top of the tables and various periods, called "Years," along the left side of the tables. You find the entry for the rate and period you are looking for in the top and left side entries. The entry in the intersection is the value you are looking for. Note that in these tables, the time column is called "Years." Actually, in our usage, "periods" would be a better title.

Look at the table headed "Amount at Compound Interest." This is the amount at compound interest, also called the future value of 1. If you have a different amount on deposit at the beginning, you must multiply this amount by the entry shown in the table to compute the total future value.

For example, suppose you want to find the amount your 8% deposit has accumulated to after 2 years. You look at the table for compound interest, for an 8% rate, and for Years = 2. At the intersection, you find the entry 1.16640000. This is also the value we computed earlier, and we hoped and expected that the table would agree with our calculation. Multiply this by $100, your original deposit, to calculate the total amount on deposit.

EXAMPLE 3.1.   You deposit $1,000 in the bank for 4 years at 6% true annual interest, compounded annually. How much will you have at the end of 4 years?

Look in the table for $i = .06$, and n (years or periods) = 4. (Note: In this Chapter we will use t for the number of years and n for the number of compoundings within 1 year.) You find the entry 1.26247696. This will be the value of 1 at 6% in 4 years. Multiply this value by 1,000 and you get 1,262.47696. This will be the value of your deposit in 4 years.   ∎

EXAMPLE 3.2.   You deposit $2,954.37 in the bank at 5% true annual interest, compounded annually. How much will you have at the end of 17 years?

Look in the table for $i = .05$ and n = 17. You find the entry 2.29201832. This will be the value of 1 at 5% in 17 years. Multiply this number by 2,954.37 (your original deposit), and you get $6,771.47. This will be the value of your deposit in 17 years.

It is possible for the compounding also to apply to noncash transactions. Here is an example.   ∎

EXAMPLE 3.3   Remember the lakeside cottage example from Chapter 2. Your neighbor, with your help, finished the cottage. However, during the first summer, with celebrations for its completion, visiting relatives, and more than expected rental income, he has not been able to let you use it. He proposes that you forego use during the first summer. During the second summer, he will let you use the cottage for 5 weeks, as follows:

2 weeks: Your planned use for the second summer

2 weeks: Your planned use for first summer, postponed to the second summer

1 week : Your compensation for not using the cottage during the first summer

You have earned an extra week of use, based on not using your week during the first summer; you have "invested" the 2 weeks for 1 year and "earned" an extra week of use. You can think of your extra week during the second summer as "interest on interest" on the 2 weeks you had to postpone during the first summer. This is the equivalent of 50% per year return.

The return does not need to be in the same form as the investment. Here is an example.   ■

EXAMPLE 3.4.   Your neighbor suggests that you postpone the first summer's use until the second summer, and he will pay you $100 for this postponement. You have earned $100 as "interest on interest" for your 1-year "investment" of your 2 weeks of use.

If you value each week of use at an assumed $200 rental value, your "interest on interest" has earned at a 25% annual rate (if he pays you the next summer). Only you can say whether this is more advantageous than taking the extra week with a value of $200. Perhaps you would rather have the extra week than the $100 cash.   ■

LOOKING AT THE COMPOUND INTEREST TABLES

Look at any column in the compound interest tables. These will be the future value of 1 for a particular interest rate. You can see that the values increase as you go down the column. They increase because the number of periods increases. Money lent out for a longer time accumulates to more money.

Look at any row across the columns for a number of interest rates. The values increase as you go from left to right. They increase because the interest rates increase as you go from left to right. Money invested at a higher rate accumulates to more money in an equal time.

When you look at the tables, you can get an idea of the values for a variety of rates and periods that you cannot get by simply using a computer.

COMPOUNDING WITHIN A PERIOD

Suppose the bank still pays you at a rate of 8% per year, but pays semiannually. The flow of funds looks like this:

Start	100.00
First half year interest	4.00
At end of half year	104.00
Second half year interest	4.16
Total at year end	108.16

At the end of the year, you have earned 8.16 in interest. Compounding within the year has earned you an extra $.16 of interest. The bank has paid you a nominal annual rate of 8%, compounded semiannually. You can see that this is the same as a 4% rate paid semiannually.

Suppose your bank pays 8% annually, but compounded quarterly. The flow of funds looks like this:

Start	100.00
First quarter interest	2.00
End of first quarter	102.00
Second quarter interest	2.04
End of first half	104.04
Third quarter interest	2.0808
End of third quarter	106.1208
Fourth quarter interest	2.122416
End of year	108.243216

At the end of the year, you have earned 8.2432 in interest. The bank has paid you at a rate of 8% per year, compounded quarterly. This is the same as a rate of 2% per quarter.

If the bank pays you 8% annually, compounded monthly, you would have 108.29995068 on deposit. This is the same as a rate of (2/3)% per month. You earned 8.29995 in interest during the year.

Suppose the bank pays you 8% annually, compounded daily. How much would you have at the end of the year, assuming a 360-day year? You would have earned 8.3277440 in interest, and have 108.3277440. If you assume a 365-day year, you will have 108.3277572 on deposit. This is slightly more than you would have assuming a 360-day year.

THE EQUATIONS FOR COMPOUND INTEREST– COMPOUNDING WITHIN A PERIOD

Equation 3.3 shows that the total amount after t years, at interest rate i, was

$$S_t = S(1+i)^t \qquad \text{Equation 3.3}$$

For compounding within a period, we divided i by the number of compounding periods within the period, and multiplied the exponent by the same number of compounding periods. For example, for quarterly compounding, we divided the interest rate of 8% by 4 (quarters in 1 year) and multiplied the period (1 year, in this case) by 4. Our equation then became

$$S_4 = S(1 + .8/4)^4 \qquad \text{Equation 3.4}$$

For t years, we have

$$S_{4t} = S(1 + .8/4)^{4t} \qquad \text{Equation 3.5}$$

We can generalize this equation, as follows:

> Suppose S is lent out at
> i% per year, expressed as a decimal,
> for t years,
> at n compoundings per year

It will then amount to the following:

$$S_{nt} = S(1 + i/n)^{nt} \qquad \text{Equation 3.6}$$

We say that S earns at a nominal annual rate of i per year, compounded n times per year. For example, if you borrow to buy a house at 8% per year with monthly payments, you have borrowed at an 8% annual rate, compounded monthly. This is the same as $(8/12)\% = (2/3)\%$ per month.

CONTINUOUS COMPOUNDING: HOW IT WORKS AND WHEN IT APPLIES

In a previous section, we showed the results of compounding more frequently within a year. Here they are again, rearranged for more clarity:

Compounding Frequency	Interest Earned
Annually	8.00
Semiannually	8.16
Quarterly	8.243216
Monthly	8.29995
Daily (360 days)	8.3277440
Daily (365 days)	8.3277572

You can see that as the compounding frequency increases, the total return for the year also increases. However, the rate of increase of the total return decreases.

What happens if we have compounding instantaneously (or momently, or continuously) at all times? Do we reach a peak amount of interest earned?

Yes, we do. This is called continuous compounding, and compound interest functions with continuous compounding are called continuous compound interest functions. Your original investment earns interest at every moment, and the interest in turn earns interest every moment, starting the moment the interest itself is earned.

THE DERIVATION OF THE EQUATIONS FOR CONTINUOUS COMPOUNDING

Here are three different approaches to figuring the future value with continuous compounding. You should use the approach you are most comfortable with. This will depend on your mathematical skills. However, no matter what your math background, you should have some understanding of how continuous functions work, when you might use them, and when and how to apply the equations.

If you have never studied calculus, read Approach 1. It will give you an intuitive understanding of continuous functions, how they are developed, and what they mean. If you have studied calculus, read Approach 2. It will develop the relationship between continuous compounding and the mathematical constant e. You will understand better what is going on with continuous compounding. If you have studied differential equations (most people have not), read Approach 3 (as well as Approach 2). This will give you a wider vision of how interest formulas might be developed.

APPROACH 1

Here is a table showing compounding periods for the 8% bank deposit we discussed earlier, with some additions for more frequent compounding.

Compounding Frequency	Interest Earned
Annually	8.00
Semiannually	8.16
Quarterly	8.243216
Monthly	8.2999507
Daily (360 days)	8.3277440
Daily (365 days)	8.3277572
1,000 times	8.3283601
10,000 times	8.3286721

You can see that the amount earned increases as the number of compoundings increases (and the compounding period decreases), but it increases more slowly. You might ask whether it reaches some highest level.

Mathematicians call such a set of numbers an "infinite series," and the individual members of the series are called the "terms" of the series. Mathematicians ask, "Does the series converge?" This means, in a general way, does the series approach some number, and after enough terms, stay within some arbitrarily small range, picked by you, around that number?

This series does converge, and it converges to the number 8.3287068, to seven decimal places. How do we compute it? For that series the answer is $100(e^{.08} - 1)$, where e is the base of the natural logarithms.

Almost all the equations you will encounter in this book will approach limits and have derivatives at all except a relatively few points.

You can see that you can come pretty close to the continuous function by compounding a reasonable number of times. For example, daily compounding (either 360 or 365 days) puts you within .001 of the correct answer in the preceding case. This is probably close enough for most practical purposes. If you want to get closer, compound more frequently. In the previous case, compounding 10,000 times per year brings you within .00005 of the correct answer. How close do you need to get? Only you can answer that, and the answer depends on your needs.

APPROACH 2

The future value of 1 at compound interest for rate i and time (in periods, such as years) t is computed as follows:

$$(1+i)^t$$

For compounding n times, the equation is

$$\text{Future value of } 1 = \left(1 + \frac{i}{n}\right)^{nt}$$

Equation 3.7

Remember the definition of e:

$$e = \frac{\lim}{n \to \infty}\left(1 + \frac{1}{n}\right)^n$$

Equation 3.8

(Note that ∞ is the mathematical symbol for infinity.)

Multiplying numerator and denominator of the second term of equation 3.7 by $1/i$, and multiplying the exponent by i/i ($i \neq 0$), we have

$$\text{Future value of } 1 = \left(1 + \frac{1}{\frac{n}{i}}\right)^{\left(\frac{n}{i}\right)it}$$

Equation 3.9

Taking the limit as n approaches ∞, we have

$$\lim_{n \to \infty} \left(1+\frac{1}{\frac{n}{i}}\right)^{\left(\frac{n}{i}\right)it} = e^{it} \qquad \text{Equation 3.10}$$

Note that if $i = 0$, then the amount at compound interest is always 1.

APPROACH 3

During time dt, we may think of principal S as earning an amount dS at interest rate i. We develop the differential equation to describe this.

$$dS = i\,S\,dt \qquad \text{Equation 3.11}$$

This differential equation has the general solution

$$S = ce^{it}, \text{ where } c \text{ is a constant} \qquad \text{Equation 3.12}$$

If $S = 1$ at time $t = 0$, then we have

$$1 = ce^{0i}, \text{ or} \qquad \text{Equation 3.13}$$

$$1 = c \times 1, \text{ or}$$
$$c = 1 \qquad \text{Equation 3.14}$$

We have in this case

$$\text{Compound value } = e^{it} \qquad \text{Equation 3.15}$$

This is the same equation as developed in Approaches 1 and 2, as we would expect. However, this approach can possibly be used when the earnings equation is more complicated than a percentage earned per period. It might vary by time, or by amount, or have a probability component. We'll discuss some of these possibilities later in the book.

WHAT IS A MATHEMATICAL MODEL?

Many readers might not have ever worked with mathematical models and therefore might not have a clear understanding of what we mean by the term. This section tries to explain the concept.

Many of you made models when you were young, possibly plane, boat, or train models, or perhaps you played with model trains. These models were small representations of a real plane, boat, or train. They showed only the main features of the thing being modeled. With larger size, or more work in modeling, they could resemble the original more accurately, but no one would ever confuse them with the original. For example, ink lines might mark the flaps and windows in a plane model. Larger models might have the rivets depicted as well. Some plane models even had working propellers, powered by a small engine or perhaps a rubber band, so the plane would fly, after a fashion. But no matter how elaborate the model was, it was only a model, only a representation of the thing depicted, not the original. At the same time, each model was meant to represent a particular thing, such as a particular railroad locomotive on a particular railroad line, or a more general representation, such as a Lionel Lines general model of a steam locomotive, which might differ from any specific real locomotive.

Models have had many practical uses. For many years, the United States Patent Office required a working model to be submitted with each patent application. It has now sold most of these models, but they are in many collections, in museums, and privately held. Aircraft manufacturers have tested models of proposed new planes in wind tunnels to observe how the plane might actually fly.

Mathematical models perform much the same function. They represent, in mathematical equation form, some activity or object. We use them because we can apply some general theory to learn more about the activity or object described by the equations, or to predict the behavior of the activity or object.

SOME FAMOUS MATHEMATICAL MODELS

One of the most famous developments of mathematical models is the development of equations for the movements of the planets. Copernicus is generally credited with first developing the idea, in the 16th century, that the planets moved in circular orbits around the sun. The mathematical model of this behavior would be a circle, with the sun at the center. In the early 17th century, Kepler improved the model. He showed that the planets moved around the sun in elliptical orbits; a circle is a special case of an ellipse. Later in the 17th century, Sir Isaac Newton developed his laws of motion and used them to describe the movements of the planets. However, these equations did not predict the movements perfectly. Some irregularities, called perturbations by astronomers, existed. These led astronomers in the 19th century to discover the planet Neptune. You cannot see Neptune with the naked eye, but you can see it using a telescope if you know where to look. An analysis of the perturbations showed astronomers where to look. We now have mathematical models of the movements of the heavenly bodies that are quite well refined, with centuries of observations to draw on, as well as the reports sent back by more recent interplanetary rockets. You can see how the models became better over time, much as if additional ink markings for rivets were added to a plane or train model. Yet Kepler's original equations describe the planetary movements quite well and are, by themselves, a good model.

But if you want to develop atom bombs or nuclear power, you must use a somewhat different set of models (or equations), developed by physicists during the 20th century. These include theories and equations about atomic behavior.

We used two mathematical models earlier in this chapter when we developed equations for compound interest and for continuous compounding. We did not present them as models, though, just as equations. However, you can use the both equations to model the growth of a savings account or an investment account, such as an IRA or a 401(k) plan. You simply make an assumption about the periodic earning rate and compute the future value.

REASONS FOR USING CONTINUOUS
FUNCTIONS IN FINANCIAL MODELS

Why would anyone want to use continuous functions? They are used because they represent some business activities more accurately and are much easier to use in these cases. Business and other operations that can be represented by a mathematical model, and that have continuous flows of income and expenses, may find it useful to use continuous functions in modeling their activities' functions. A business that has investment income flowing in every day, makes investments every day, has business revenues and expenses accruing every day, and has a business requirement for reserves might find continuous functions useful in modeling its activities.

A BUSINESS EXAMPLE OF USE OF
CONTINUOUS FUNCTIONS

The life insurance industry receives insurance premiums and investment income, pays claims, makes investments, and has operating expenses. Many life insurance companies have used continuous functions for many years. The author worked for one large life insurance company that started using continuous functions in 1947. Other types of insurance companies could do that as well; some casualty insurance companies also use continuous functions.

Of course, the actual work of an insurance company is done in discrete amounts. Investments are made individually, individual investment income amounts are received during each day, and these amounts will usually vary from day to day. Premiums are received in varying amounts day by day, and claims are made and expenses incurred in varying amounts day by day. However, as a model, continuous functions might easily be an excellent approximation of the actual operations and might be much easier to use.

The life insurance industry needs these models to calculate premiums and policy reserves. These calculations require projections of future values, which in turn require assumptions about future interest rates, future mortality rates, and future insurance company costs. These must be put in mathematical form. Continuous functions make the job of creating these mathematical models much easier for the insurance company and for the entire insurance industry. People who work extensively with these models are called actuaries. We'll discuss some of these models later in the book.

FURTHER REFLECTIONS ON APPROACH 3

In Approach 3, we found a function for S based on the solution of a differential equation. The exponent of e was simply the expression (it), with i a constant. Actually, the exponent could be any function of t, as well as other variables, although this would (usually) result in a different differential equation. The equation for S could actually be any function of t; it need not be an exponential. In this chapter, we use a constant coefficient, called i. Any equation is possible in theory. In the real world of finance, only a relatively few equations could be realistically used.

Later in the book, we will discuss yield curves and develop a spot rate. This could lead to a variety of approximations for an equation for interest rates.

COMPUTING i, GIVEN S, S_{NT}, T, AND N

Suppose you are given a final value of a compounding, together with the initial value, the time, and the number of compoundings per year. How would you compute the interest rate that gave the final value? Here are several methods. S and S_{nt} are the beginning and ending value, respectively, of the amount.

METHOD 1: THE ALGEBRAIC SOLUTION

In this section, we solve the general equation for future value for i, given the beginning investment (S), the final investment (S_{nt}), and t and n. We will do the case of one compounding per period first.

Suppose an original investment of S is now worth S_t after t years. The equation is

$$S_t = S(1+i)^t$$ Equation 3.16

Dividing through by S, we have

$$(S_t/S) = (1+i)^t$$ Equation 3.17

This converts the equation to a compound value of 1.
Taking the tth root of each side, we have

$$(S_t/S)^{1/t} = 1+i$$ Equation 3.18

And therefore

$$i = (S_t/S)^{1/t} - 1$$ Equation 3.19

Now suppose we compound n times per year, but still for t years. The new equation is

$$S_{nt} = S(1+i/n)^{nt}$$ Equation 3.20

Dividing through by S, as before, we have

$$(S_{nt}/S) = (1+i/n)^{nt}$$ Equation 3.21

Taking the $(nt)^{th}$ root of each side, we have

$$(S_{nt}/S)^{1/(nt)} = (1+i/n)$$ Equation 3.22

Subtracting 1 from both sides, and multiplying each resulting side by n, we have

$$i = n\left[(S_{nt}/S)^{1/nt} - 1\right]$$ Equation 3.23

METHOD 2: CALCULATING i, GIVEN S, S_{nt}, T, AND N (A NUMERICAL ANALYSIS APPROACH)

Suppose you are given S, S_{nt}, n, and t, and you wish to know the value of i, which gave S_{nt}, starting with S. How would you do it?

Many calculators have a function that will compute this for you. Here are two other methods you can use. The first, the bisection method, can be used in many approximation problems presented in this book. You should have some idea how to use these methods.

The Bisection Method for Solving Equations: A Practical Approach

This method is not nearly so formidable as its name. In fact, one book on numerical analysis calls this method "One of the best, most effective methods for finding the zeros of a continuous function." In his 710 page book, the author devotes precisely two pages (about .3%) to this method.[1] It is that easy to learn and to use. We'll use a variant of this method to calculate the square root of 17; we will then look at the mathematics of the bisection method; and then we will use it again to find *i*, given *S*, S_{nt}, n, and t in an example. This method is one of a set of iterative techniques. It has also been called the "method of successive approximations" and sometimes wrongly called "trial and error."

Computing the square root of 17

Suppose you need to compute the square root of 17. How might you go about it? Here is a procedure.

Start with the integer 1, and continue taking squares until you reach an integer whose square is greater than 17. Then go back halfway into the interval you just covered. Square this midpoint number. Based on the result, pick either the lower or the upper subinterval. If the result is greater than 17, pick the lower subinterval; if the result is less than 17, pick the higher subinterval. Pick the midpoint of this new subinterval. Square this number. Continue this process until you decide to stop.

When do you stop? When you are close enough to meet your requirements. Only you, or your boss, can decide that. The job you are doing will determine what you need for requirements. Many calculators and computer programs automatically stop after a while. In these cases, the calculator manufacturer or computer programmer has already built in a stopping point. You might be able to change it, but it probably is not worth doing so.

To compute an approximation for the square root of 17, we start with 1 squared = 1, less than 17, so

$$
\begin{aligned}
2 \quad \text{squared} &= 4, \text{ less than 17, so} \\
3 \quad \text{squared} &= 9, \text{ and} \\
4 \quad \text{squared} &= 16, \text{ and} \\
5 \quad \text{squared} &= 25
\end{aligned}
$$

Now we have our first interval, (4, 5). We now know that

$$4 < \text{square root of } 17 < 5$$

If we set the square root of 17 = 4.5, we know we are within .5 of the true number. This might be close enough, but it probably is not, so we continue.

---

[1] The quotation above is from Hamming, p. 62. The mathematical development is based on Conte and de Boor, slightly modified.

To continue, we bisect the interval (4, 5) and use the bisection number, 4.5

$$4.5 \text{ squared} = 20.25, \text{ bigger than } 17$$

We must choose the new interval (4, 4.5), because the square root of 17 must be less than 4.5. We bisect this new interval, and square the midpoint number

$$4.25 \text{ squared} = 18.0625$$

This is greater than 17, but we're getting closer. We bisect again, using the new number 4.125

$$4.125 \text{ squared} = 17.015625$$

You decide to pick 4.0625 for your approximation, because you know that it is within .0625 of the correct answer and is close enough for your purposes. My calculator gives 4.123105626 as the answer. This is within .0625 of the approximation, although an earlier approximation, 4.125, is actually closer. This can happen with the bisection method.

You can see why this method, and similar methods, can be called "successive approximations." You continue to approximate the answer, getting closer and closer to it. If the equation has a solution that is a rational number (a number that can be expressed as a fraction of two integers), then this method can produce the exact answer. The square root of 17 is not rational, so you can only get an approximation using this or any other method. If the equation to be solved was

$$x^2 = 20.25$$

you would get a precise answer soon, because the square root of 20.25 is 4.5. You would get that answer on the first bisection.

You can also see why the name "trial and error" is wrong, at least to me. Trial and error implies trying different numbers at random. With the bisection method, you know exactly what numbers to try next, and you know exactly when to stop.

The Bisection Method: A Mathematical View

This section presents the bisection method from a more mathematical point of view. We'll use the example (finding the square root of 17) in the previous section to illustrate the points.

We wish to find the solutions of an equation. We first put this equation in the form

$$f(x) = 0$$

Mathematicians call the solutions to this the "zeros" of the equation.

In our example, the equation would be

$$f(x) = x^2 - 17 = 0$$

and we wish to find the zeros of this equation. Actually we can solve this equation easily by factoring it into the form [x + Sqrt(17)][x − Sqrt(17)] = 0, but most equations cannot be solved this easily.

We need to find an interval $[a_0, b_0]$ in which

1. the function is continuous, and
2. the function has a zero.

The test for this condition is that $f(a_0)f(b_0) < 0$.

We found the interval [4,5], which met this condition, because f(4) = (16 − 17) = −1 < 0, and f(5) = (25 − 17) = 8 > 0, and (−1)(8) = −8 < 0. Our example is continuous in the interval (it is actually continuous everywhere), so we know the interval contains a zero.

$$\text{For } n = 0, 1, 2, \ldots,$$
$$\text{set } m = (a_n + b_n)/2$$

If $f(a_n)f(m) < 0$, there is a sign change in the interval $[a_n, m]$. Set $a_{n+1} = a_n$, and $b_{n+1} = m$.

If $f(a_n)f(m) > 0$, there is a sign change in the interval $[m, b_n]$. Set $a_{n+1} = m$, and $b_{n+1} = b_n$.

If $f(a_n)f(m) = 0$, then m is a zero. This happened earlier when we set $x^2 = 20.25$ and got our zero on the very first bisection.

When you select an interval with a sign change, the interval could have more than one zero. If you pick large intervals, you have a higher chance of getting an interval with multiple zeros. If you pick a small interval size, you will spend considerable time searching where there will be no zeros. In any case, be sure to check on whether the interval you have selected contains multiple zeros. This can be troublesome to check by computer.

Relatively few bisections are needed for most fixed income math applications. Ten bisections produce an improvement of 1/1,000, or a better than three decimal place accuracy in the answer. Few financial mathematics applications need such accuracy.

Professional mathematicians will comment, correctly, that there are many much more powerful algorithms than the bisection method. That is true. However, the bisection method is easy to understand and to use. You can explain it to your boss if you have to. You can use it with calculators. Study of the more powerful algorithms may more properly belong to a course on numerical analysis, not fixed-income mathematics.

ACCURACY REQUIREMENTS

You set the accuracy requirements for your approximation. This really means that the job you are doing sets the requirements. Different jobs will have different requirements. For example, most financial mathematics approximation jobs do not require particularly great precision. Making rockets to circle the planet Jupiter will have great precision requirements. Determining these precision requirements might be the hardest part of the actual approximation job.

LEGAL REQUIREMENTS FOR ACCURACY

Sometimes the accuracy requirements are stated legally. For example, the Municipal Securities Rulemaking Board (MSRB), one of the regulatory authorities for the municipal securities industry, requires that the yield equivalent of the dollar price must be shown on the customer confirmation, accurate to within .05% of yield, if the bond was traded at a dollar price. The MSRB has devoted a whole section of a rule to accuracy requirements. You should always make sure that any legal or regulatory requirements have been met in calculation of approximate numbers. Check with the rule publications, or your compliance officer, for the requirements. A book like this is not a good source, because the rules are subject to change.

AN EXAMPLE FROM COMPOUND INTEREST

Suppose that after 10 years, a $100 investment has grown to $160. What was the annual return?
    We need to find an interval that contains the 160 final value.

We try 3%, and find the value is 134.39, so
We try 4%, and find the value is 148.02, so
We try 5%, and find the value is 162.08, and we have our desired
    interval, [4, 5].
We bisect the interval, and
Try 4.5%, and find the value is 155.30, so we bisect the interval [4.5, 5],
    and
Try 4.75%, and find the value is 159.05, so we bisect the interval [4.75,
    5], and
Try 4.875, and find the value is 160.96,
We decide we are close enough, so we bisect the interval [4.75, 4.875]
    and obtain 4.8125 as the approximate interest rate. We are within
    .0625 of the correct answer.

USING TABLES AND INTERPOLATING BETWEEN VALUES

Sometimes you can use the tables to approximate the answer, using a linear interpolation. For example, in the preceding case, you can find the compound values of 4.5% and 5% as follows:

4.5%	1.5530
5%	1.6289

We assume that the relative distance between 4.5% and the correct answer is the same as the distance between 1.5530 and 1.60. Let x be the desired interest rate i. Then we have

$$(x-4.5)/(5.0-4.5) = (1.60-1.5530)/(1.6289-1.5530), \text{ or}$$
$$(x-4.5)/.5 = (.0470)/(.0759), \text{ or}$$
$$(x-4.5) = (.5)(.6192) = .30962, \text{ and}$$
$$x = 4.80962$$

This compares with our answer of 4.8125 using the bisection method. The two answers are the same to two decimal places. Financial mathematics rarely requires more accuracy.

THE RULE OF 72

A well-known rule, the "Rule of 72," states that the number of years required for money to double, at interest rate i, equals 72 divided by the interest rate. For example, money at 6% doubles in about 12 (= 72/6) years. This gives fairly good results for a wide range of interest rates.

A ZERO INTEREST RATE?

An interest rate of zero is certainly mathematically possible. What would it mean in economic terms? It would mean that any venture or activity that would eventually earn the amount spent would be worth doing, even if it did not reward the capital at all. This seems unlikely. Although in the past interest rates have been quite low at times, a zero interest rate seems unlikely. I certainly have never heard of one. However, during World War II, the United States Treasury pegged the yield on 90-day bills at .375%, so that $1 million in bills would yield $3,750 annually. The reward hardly seems worth the work to get it.

In late April 2001, several companies actually sold bonds with a zero interest rate. However, these were convertible bonds. The volatility and an attractive conversion ratio made this an attractive purchase to many buyers. In this case, the value of the conversion privilege accounted for the zero coupon rate.

NEGATIVE INTEREST RATES?

Negative interest rates are also mathematically possible. A negative interest rate would mean that the lender would actually pay borrowers for the privilege of lending them money. It hardly seems likely that this would ever occur. However, years ago, maturing U.S. Treasury securities sometimes sold at negative yields. At that time, the Treasury frequently sold new securities by offering them to holders of maturing securities. The rights to purchase the new securities sometimes had a value high enough to make the yield to maturity of the maturing securities negative. Needless to say, this reflected the value of the rights, not the cost of money. The companies mentioned earlier, with zero-interest-rate convertible bonds, might have been able to obtain a negative interest rate if they made the conversion feature attractive enough. I have never heard of an actual negative interest rate in a normal lending transaction. However, it is certainly possible for service charges and fees to cost more than the interest earned on an investment. For example, a small savings account might have more in charges than it earned in interest.

REAL AND NOMINAL RATES

Suppose you lend some money at 6%, and in one year you receive your 6% interest payment. You have received the stated rate on your loan. This is also called the nominal rate of interest.

But suppose during the year the money has decreased in value, according to any reasonable measure of value. In the United States, this could be the consumer price index (CPI). In terms of real purchasing power, you have not earned 6%, but somewhat less. For example, suppose the CPI shows an inflation of 5% during the year. You have earned a nominal rate of 6%, but after inflation you have earned only 1% (6% − 5%). This is called the real rate of return.

Real rates of return can easily be negative. For example, if inflation was 7% during the years, you would have a negative 1% real rate of return. Real returns were frequently negative during inflationary times in the 20th century.

You should know the meaning of nominal and real rates of return. However, an analysis of real returns is more properly a part of economics rather than fixed-income mathematics.

CHAPTER SUMMARY

The equation for the amount at compound interest is

$$(1 + i/n)^{nt}$$

Where i = interest rate, expressed as a decimal
t = number of periods
n = number of compoundings per period

The process of interest earning interest is called compounding. The interest earned by the interest is called "interest on interest."

To compute the actual amount you have at the end of the interest time, multiply the starting amount by the compound value of 1.

Continuous compounding consists of interest paid instantaneously and invested instantaneously.

The equation for the compound value of 1 at continuous interest is

$$e^{it}$$

We showed three ways of looking at continuous compounding: (1) compounding more and more frequently, (2) developing the equation involving e by taking the compounding process to a limit, and (3) solving the differential equation:

$$dS = i\, S\, dt$$

A mathematical model is a representation of an activity or object by a mathematical equation.

Continuous interest functions can be used in a variety of business applications, including insurance.

Given S, S_{nt}, t, and n, the solution for i is

$$i = n\left[(S_{nt}/S)^{1/nt} - 1\right]$$

The bisection method works as follows:

1. set an interval that contains a zero of your equation,
2. determine the midpoint of the interval,
3. compute the value of the function at the midpoint,
4. based on your answer, select either the upper or lower half of your first interval as the new interval, and
5. continue until you are satisfied with the result.

Be sure to consider your need for accuracy requirements. Sometimes the requirements are set by law, regulation, trade custom, or your boss.

Sometimes it is relatively easy to do a linear interpolation using interest tables.

The Rule of 72 states that the time required for money at compound interest to double approximately equals 72 divided by the interest rate.

COMPUTER PROJECTS

1. Make a compound interest table showing the values for a range of interest rates and periods. Do you get the same answers as those in the tables in the book?
2. Make a chart of the time it would take for an investment to double, using a wide range of interest rates, say from .01% to 100% annually. How do the results differ from the results predicted by the Rule of 72? For what range of interest rates would you consider the Rule of 72 a worthwhile approximation?

TOPICS FOR CLASS DISCUSSION

1. In 1624, according to tradition, the Dutch paid the indians $24 in trade goods for Manhattan Island. Apply any reasonable interest rate, say 6%. Was this a good deal for the Dutch, considering the present value of Manhattan Island? What other factors would you consider in analyzing this problem?
2. Take any reasonable sum that a rich man might have in the year 1 AD, say $100,000 in gold. Apply any reasonable interest rate, say 6%. How does this compare to the present value of all the property in the world? What does this say about the earning power of capital, the cost of capital destruction (say by war, accident, or natural disaster), and the risk of capital obsolescence?
3. What is so magical about the number 72? Why should the Rule of 72 work?

PROBLEMS

1. You have a $10,000 loan from the bank. You have been paying 8% per year, with semiannual payments of interest. The bank now wants you to pay interest monthly. What was your old true annual rate and your new true annual rate?
2. In settling your father's estate, you have come across an old certificate of deposit, dating from 1955, with an original deposit of $1,000 and interest at 3% annually. How much is it worth now?

3. You contacted the bank, which told you that starting in 1960, it credited interest quarterly instead of annually. How much does this add to the value of the CD?
4. What is the equation for the compound value of 1, at interest rate i, for t years, compounded n times per year?
5. Your mother bought a U.S. savings bond in 1943, paying $18.75 for a bond that matured at $25 10 years later. What rate did she earn?

SUGGESTIONS FOR FURTHER STUDY

For differential equations, the following is a well-known and widely used text: Boyce, William E., and DiPrima, Richard C., *Elementary Differential Equations and boundary Value Problems,* 6th ed., John Wiley & Sons, New York, 1997.

For numerical analysis, the author has used: Conte, S. D., and de Boor, Carl, *Elementary Numerical Analysis: An Algorithmic Approach,* 3rd ed., McGraw-Hill, New York, 1980.

Hamming, R. W., *Numerical Methods for Scientists and Engineers,* 2nd ed., Dover, New York, 1986.

AMOUNT AT COMPOUND INTEREST $(1 + i)^n$

The following table gives the amount after a term of n years on unit original principal at rate of interest i.

Years n	Rate i				
	.0025($\frac{1}{4}$ %)	.004167($\frac{5}{12}$ %)	.005($\frac{1}{2}$ %)	.005833($\frac{7}{12}$ %)	.0075($\frac{3}{4}$ %)
1	1.00250000	1.00416667	1.00500000	1.00583333	1.00750000
2	1.00500625	1.00835069	1.01002500	1.01170069	1.01505625
3	1.00751877	1.01255216	1.01507513	1.01760228	1.02266917
4	1.01003756	1.01677112	1.02015050	1.02353830	1.03033919
5	1.01256266	1.02100767	1.02525125	1.02950894	1.03806673
6	1.01509406	1.02526187	1.03037751	1.03551440	1.04585224
7	1.01763180	1.02953379	1.03552940	1.04155490	1.05369613
8	1.02017588	1.03382352	1.04070704	1.04763064	1.06159885
9	1.02272632	1.03813111	1.04591058	1.05374182	1.06956084
10	1.02528313	1.04245666	1.05114013	1.05988865	1.07758255
11	1.02784634	1.04680023	1.05639583	1.06607133	1.08566441
12	1.03041596	1.05116190	1.06167781	1.07229008	1.09380690
13	1.03299200	1.05554174	1.06698620	1.07854511	1.10201045
14	1.03557448	1.05993983	1.07232113	1.08483662	1.11027553
15	1.03816341	1.06435625	1.07768274	1.09116483	1.11860259
16	1.04075882	1.06879106	1.08307115	1.09752996	1.12699211
17	1.04336072	1.07324436	1.08848651	1.10393222	1.13544455
18	1.04596912	1.07771621	1.09392894	1.11037182	1.14396039
19	1.04858404	1.08220670	1.09939858	1.11684899	1.15254009
20	1.05120550	1.08671589	1.10489558	1.12336395	1.16118414
21	1.05383352	1.09124387	1.11042006	1.12991690	1.16989302
22	1.05646810	1.09579072	1.11597216	1.13650808	1.17866722
23	1.05910927	1.10035652	1.12155202	1.14313771	1.18750723
24	1.06175704	1.10494134	1.12715978	1.14980602	1.19641353
25	1.06441144	1.10954526	1.13279558	1.15651322	1.20538663
26	1.06707247	1.11416836	1.13845955	1.16325955	1.21442703
27	1.06974015	1.11881073	1.14415185	1.17004523	1.22353523
28	1.07241450	1.12347244	1.14987261	1.17687049	1.23271175
29	1.07509553	1.12815358	1.15562197	1.18373557	1.24195709
30	1.07778327	1.13285422	1.16140008	1.19064069	1.25127176
31	1.08047773	1.13757444	1.16720708	1.19758610	1.26065630
32	1.08317892	1.14231434	1.17304312	1.20457202	1.27011122
33	1.08588687	1.14707398	1.17890833	1.21159869	1.27963706
34	1.08860159	1.15185346	1.18480288	1.21866634	1.28923434
35	1.09132309	1.15665284	1.19072689	1.22577523	1.29890359
36	1.09405140	1.16147223	1.19668052	1.23292559	1.30864537
37	1.09678653	1.16631170	1.20266393	1.24011765	1.31846021
38	1.09952850	1.17117133	1.20867725	1.24735167	1.32834866
39	1.10227732	1.17605121	1.21472063	1.25462789	1.33831128
40	1.10503301	1.18095142	1.22079424	1.26194655	1.34834861
41	1.10779559	1.18587206	1.22689821	1.26930791	1.35846123
42	1.11056508	1.19081319	1.23303270	1.27671220	1.36864969
43	1.11334149	1.19577491	1.23919786	1.28415969	1.37891456
44	1.11612485	1.20075731	1.24539385	1.29165062	1.38925642
45	1.11891516	1.20576046	1.25162082	1.29918525	1.39967584
46	1.12171245	1.21078446	1.25787892	1.30676383	1.41017341
47	1.12451673	1.21582940	1.26416832	1.31438662	1.42074971
48	1.12732802	1.22089536	1.27048916	1.32205388	1.43140533
49	1.13014634	1.22598242	1.27684161	1.32976586	1.44214087
50	1.13297171	1.23109068	1.28322581	1.33752283	1.45295693

Reprinted with permission. From *Mathematical Tables from Handbook of Chemistry & Physics, Tenth Edition.* Copyright CRC Press, Boca Raton, Florida.

AMOUNT AT COMPOUND INTEREST $(1 + i)^n$
(Continued)

Years n	.0025($\frac{1}{4}$ %)	.004167($\frac{5}{12}$ %)	.005($\frac{1}{2}$ %)	.005833($\frac{7}{12}$ %)	.0075($\frac{3}{4}$ %)
			Rate i		
50	1.13297171	1.23109068	1.28322581	1.33752283	1.45295693
51	1.13580414	1.23622022	1.28964194	1.34532504	1.46385411
52	1.13864365	1.24137114	1.29609015	1.35317277	1.47483301
53	1.14149026	1.24654352	1.30257060	1.36106628	1.48589426
54	1.14434398	1.25173745	1.30908346	1.36900583	1.49703847
55	1.14720484	1.25695302	1.31562887	1.37699170	1.50826626
56	1.15007285	1.26219033	1.32220702	1.38502415	1.51957825
57	1.15294804	1.26744946	1.32881805	1.39310346	1.53097509
58	1.15583041	1.27273050	1.33546214	1.40122990	1.54245740
59	1.15871998	1.27803354	1.34213946	1.40940374	1.55402583
60	1.16161678	1.28335868	1.34885015	1.41762526	1.56568103
61	1.16452082	1.28870601	1.35559440	1.42589474	1.57742363
62	1.16743213	1.29407561	1.36237238	1.43421246	1.58925431
63	1.17035071	1.29946760	1.36918424	1.44257870	1.60117372
64	1.17327658	1.30488204	1.37603016	1.45099374	1.61318252
65	1.17620977	1.31031905	1.38291031	1.45945787	1.62528139
66	1.17915030	1.31577872	1.38982486	1.46797138	1.63747100
67	1.18209817	1.32126113	1.39677399	1.47653454	1.64975203
68	1.18505342	1.32676638	1.40375785	1.48514766	1.66212517
69	1.18801605	1.33229458	1.41077664	1.49381102	1.67459111
70	1.19098609	1.33784580	1.41783053	1.50252492	1.68715055
71	1.19396356	1.34342016	1.42491968	1.51128965	1.69980418
72	1.19694847	1.34901774	1.43204428	1.52010550	1.71255271
73	1.19994084	1.35463865	1.43920450	1.52897279	1.72539685
74	1.20294069	1.36028298	1.44640052	1.53789179	1.73833733
75	1.20594804	1.36595082	1.45363252	1.54686283	1.75137486
76	1.20896291	1.37164229	1.46090069	1.55588620	1.76451017
77	1.21198532	1.37735746	1.46820519	1.56496220	1.77774400
78	1.21501528	1.38309645	1.47554622	1.57409115	1.79107708
79	1.21805282	1.38885935	1.48292395	1.58327334	1.80451015
80	1.22109795	1.39464627	1.49033857	1.59250910	1.81804398
81	1.22415070	1.40045729	1.49779026	1.60179874	1.83167931
82	1.22721108	1.40629253	1.50527921	1.61114257	1.84541691
83	1.23027910	1.41215209	1.51280561	1.62054090	1.85925753
84	1.23335480	1.41803605	1.52036964	1.62999405	1.87320196
85	1.23643819	1.42394454	1.52797148	1.63950235	1.88725098
86	1.23952928	1.42987764	1.53561134	1.64906612	1.90140536
87	1.24262811	1.43583546	1.54328940	1.65868567	1.91566590
88	1.24573468	1.44181811	1.55100585	1.66836134	1.93003339
89	1.24884901	1.44782568	1.55876087	1.67809344	1.94450865
90	1.25197114	1.45385829	1.56655468	1.68788232	1.95909246
91	1.25510106	1.45991603	1.57438745	1.69772830	1.97378565
92	1.25823882	1.46599902	1.58225939	1.70763172	1.98858905
93	1.26138441	1.47210735	1.59017069	1.71759290	2.00350346
94	1.26453787	1.47824113	1.59812154	1.72761219	2.01852974
95	1.26769922	1.48440047	1.60611215	1.73768993	2.03366871
96	1.27086847	1.49058547	1.61414271	1.74782646	2.04892123
97	1.27404564	1.49679624	1.62221342	1.75802211	2.06428814
98	1.27723075	1.50303289	1.63032449	1.76827724	2.07977030
99	1.28042383	1.50929553	1.63847611	1.77859219	2.09536858
100	1.28362489	1.51558426	1.64666849	1.78896731	2.11108384

AMOUNT AT COMPOUND INTEREST $(1 + i)^n$
(Continued)

n	.01(1 %)	.01125(1⅛ %)	.0125(1¼ %)	.015(1½ %)	.0175(1¾ %)
			Rate i		
1	1.01000000	1.01125000	1.01250000	1.01500000	1.01750000
2	1.02010000	1.02262656	1.02515625	1.03022500	1.03530625
3	1.03030100	1.03413111	1.03797070	1.04567838	1.05342411
4	1.04060401	1.04576509	1.05094534	1.06136355	1.07185903
5	1.05101005	1.05752994	1.06408215	1.07728400	1.09061656
6	1.06152015	1.06942716	1.07738318	1.09344326	1.10970235
7	1.07213535	1.08145821	1.09085047	1.10984491	1.12912215
8	1.08285671	1.09362462	1.10448610	1.12649259	1.14888178
9	1.09368527	1.10592789	1.11829218	1.14338998	1.16898721
10	1.10462213	1.11836958	1.13227083	1.16054083	1.18944449
11	1.11566835	1.13095124	1.14642422	1.17794894	1.21025977
12	1.12682503	1.14367444	1.16075452	1.19561817	1.23143931
13	1.13809328	1.15654078	1.17526395	1.21355244	1.25298950
14	1.14947421	1.16955186	1.18995475	1.23175573	1.27491682
15	1.16096896	1.18270932	1.20482918	1.25023207	1.29722786
16	1.17257864	1.19601480	1.21988955	1.26898555	1.31992935
17	1.18430443	1.20946997	1.23513817	1.28802033	1.34302811
18	1.19614748	1.22307650	1.25057739	1.30734064	1.36653111
19	1.20810895	1.23683611	1.26620961	1.32695075	1.39044540
20	1.22019004	1.25075052	1.28203723	1.34685501	1.41477820
21	1.23239194	1.26482146	1.29806270	1.36705783	1.43953681
22	1.24471586	1.27905071	1.31428848	1.38756370	1.46472871
23	1.25716302	1.29344003	1.33071709	1.40837715	1.49036146
24	1.26973465	1.30799123	1.34735105	1.42950281	1.51644279
25	1.28243200	1.32270613	1.36419294	1.45094535	1.54298054
26	1.29525631	1.33758657	1.38124535	1.47270953	1.56998269
27	1.30820888	1.35263442	1.39851092	1.49480018	1.59745739
28	1.32129097	1.36785156	1.41599230	1.51722218	1.62541290
29	1.33450388	1.38323989	1.43369221	1.53998051	1.65385762
30	1.34784892	1.39880134	1.45161336	1.56308022	1.68280013
31	1.36132740	1.41453785	1.46975853	1.58652642	1.71224913
32	1.37494068	1.43045140	1.48813051	1.61032432	1.74221349
33	1.38869009	1.44654398	1.50673214	1.63447918	1.77270223
34	1.40257699	1.46281760	1.52556629	1.65899637	1.80372452
35	1.41660276	1.47927430	1.54463587	1.68388132	1.83528970
36	1.43076878	1.49591613	1.56394382	1.70913954	1.86740727
37	1.44507642	1.51274519	1.58349312	1.73477663	1.90008689
38	1.45952724	1.52976357	1.60328678	1.76079828	1.93333841
39	1.47412251	1.54697341	1.62332787	1.78721025	1.96717184
40	1.48886373	1.56437687	1.64361946	1.81401841	2.00159734
41	1.50375237	1.58197611	1.66416471	1.84122868	2.03662530
42	1.51878989	1.59977334	1.68496677	1.86884712	2.07226624
43	1.53397729	1.61777079	1.70602885	1.89687982	2.10853090
44	1.54931757	1.63597071	1.72735421	1.92533302	2.14543019
45	1.56481075	1.65437538	1.74894614	1.95421301	2.18297522
46	1.58045885	1.67298710	1.77080797	1.98352621	2.22117728
47	1.59626344	1.69180821	1.79294306	2.01327910	2.26004789
48	1.61222608	1.71084105	1.81535485	2.04347829	2.29959872
49	1.62834834	1.73008801	1.83804679	2.07413046	2.33984170
50	1.64463182	1.74955150	1.86102237	2.10524242	2.38078893

AMOUNT AT COMPOUND INTEREST $(1 + i)^n$
(Continued)

Years n	Rate i				
	.01(1 %)	.01125(1⅛ %)	.0125(1¼ %)	.015(1½ %)	.0175(1¾ %)
50	1.64463182	1.74955150	1.86102237	2.10524242	2.38078893
51	1.66107814	1.76923395	1.88428515	2.13682106	2.42245274
52	1.67768892	1.78913784	1.90783872	2.16887337	2.46484566
53	1.69446581	1.80926564	1.93168670	2.20140647	2.50798046
54	1.71141047	1.82961988	1.95583279	2.23442757	2.55187012
55	1.72852457	1.85020310	1.98028070	2.26794398	2.59652785
56	1.74580982	1.87101788	2.00503420	2.30196314	2.64196708
57	1.76326792	1.89206684	2.03009713	2.33649259	2.68820151
58	1.78090060	1.91335259	2.05547335	2.37153998	2.73524503
59	1.79870960	1.93487780	2.08116676	2.40711308	2.78311182
60	1.81669670	1.95664518	2.10718135	2.44321978	2.83181628
61	1.83486367	1.97865744	2.13352111	2.47986807	2.88137306
62	1.85321230	2.00091733	2.16019013	2.51706609	2.93179709
63	1.87174443	2.02342765	2.18719250	2.55482208	2.98310354
64	1.89046187	2.04619121	2.21453241	2.59314442	3.03530785
65	1.90936649	2.06921087	2.24221407	2.63204158	3.08842574
66	1.92846015	2.09248949	2.27024174	2.67152221	3.14247319
67	1.94774475	2.11602999	2.29861976	2.71159504	3.19746647
68	1.96722220	2.13983533	2.32735251	2.75226896	3.25342213
69	1.98689442	2.16390848	2.35644442	2.79355300	3.31035702
70	2.00676337	2.18825245	2.38589997	2.83545629	3.36828827
71	2.02683100	2.21287029	2.41572372	2.87798814	3.42723331
72	2.04709931	2.23776508	2.44592027	2.92115796	3.48720990
73	2.06757031	2.26293994	2.47649427	2.96497533	3.54823607
74	2.08824601	2.28839801	2.50745045	3.00944996	3.61033020
75	2.10912847	2.31414249	2.53879358	3.05459171	3.67351098
76	2.13021975	2.34017659	2.57052850	3.10041059	3.73779742
77	2.15152195	2.36650358	2.60266011	3.14691674	3.80320888
78	2.17303717	2.39312675	2.63519336	3.19412050	3.86976503
79	2.19476754	2.42004942	2.66813327	3.24203230	3.93748592
80	2.21671522	2.44727498	2.70148494	3.29066279	4.00639192
81	2.23888237	2.47480682	2.73525350	3.34002273	4.07650378
82	2.26127119	2.50264840	2.76944417	3.39012307	4.14784260
83	2.28388390	2.53080319	2.80406222	3.44097492	4.22042984
84	2.30672274	2.55927473	2.83911300	3.49258954	4.29428737
85	2.32978997	2.58806657	2.87460191	3.54497838	4.36943740
86	2.35308787	2.61718232	2.91053444	3.59815306	4.44590255
87	2.37661875	2.64662562	2.94691612	3.65212535	4.52370584
88	2.40038494	2.67640016	2.98375257	3.70690723	4.60287070
89	2.42438879	2.70650966	3.02104948	3.76251084	4.68342093
90	2.44863267	2.73695789	3.05881260	3.81894851	4.76538080
91	2.47311900	2.76774867	3.09704775	3.87623273	4.84877496
92	2.49785019	2.79883584	3.13576085	3.93437622	4.93362853
93	2.52282869	2.83037331	3.17495786	3.99339187	5.01996703
94	2.54805698	2.86221501	3.21464483	4.05329275	5.10781645
95	2.57353755	2.89441492	3.25482789	4.11409214	5.19720324
96	2.59927293	2.92697709	3.29551324	4.17580352	5.28815429
97	2.62526565	2.95990559	3.33670716	4.23844057	5.38069699
98	2.65151831	2.99320452	3.37841600	4.30201718	5.47485919
99	2.67803349	3.02687807	3.42064620	4.36654744	5.57063923
100	2.70481383	3.06093045	3.46340427	4.43204565	5.66815594

Reprinted with permission. From *Mathematical Tables from Handbook of Chemistry & Physics, Tenth Edition.* Copyright CRC Press, Boca Raton, Florida.

AMOUNT AT COMPOUND INTEREST $(1 + i)^n$
(Continued)

Years n	Rate i				
	.02(2 %)	.0225(2¼ %)	.025(2½ %)	.0275(2¾ %)	.03(3 %)
1	1.02000000	1.02250000	1.02500000	1.02750000	1.03000000
2	1.04040000	1.04550625	1.05062500	1.05575625	1.06090000
3	1.06120800	1.06903014	1.07689063	1.08478955	1.09272700
4	1.08243216	1.09308332	1.10381289	1.11462126	1.12550881
5	1.10408080	1.11767769	1.13140821	1.14527334	1.15927407
6	1.12616242	1.14282544	1.15969342	1.17676836	1.19405230
7	1.14868567	1.16853901	1.18868575	1.20912949	1.22987387
8	1.17165938	1.19483114	1.21840290	1.24238055	1.26677008
9	1.19509257	1.22171484	1.24886297	1.27654602	1.30477318
10	1.21899442	1.24920343	1.28008454	1.31165103	1.34391638
11	1.24337431	1.27731050	1.31208666	1.34772144	1.38423387
12	1.26824179	1.30604999	1.34488882	1.38478378	1.42576089
13	1.29360663	1.33543611	1.37851104	1.42286533	1.46853371
14	1.31947876	1.36548343	1.41297382	1.46199413	1.51258972
15	1.34586834	1.39620680	1.44829817	1.50219896	1.55796742
16	1.37278571	1.42762146	1.48450562	1.54350944	1.60470644
17	1.40024142	1.45974294	1.52161826	1.58595595	1.65284763
18	1.42824625	1.49258716	1.55965872	1.62956973	1.70243306
19	1.45681117	1.52617037	1.59865019	1.67438290	1.75350605
20	1.48594740	1.56050920	1.63861644	1.72042843	1.80611123
21	1.51566634	1.59562066	1.67958185	1.76774021	1.86029457
22	1.54597967	1.63152212	1.72157140	1.81635307	1.91610341
23	1.57689926	1.66823137	1.76461068	1.86630278	1.97358651
24	1.60843725	1.70576658	1.80872595	1.91762610	2.03279411
25	1.64060599	1.74414632	1.85394410	1.97036082	2.09377793
26	1.67341811	1.78338962	1.90029270	2.02454575	2.15659127
27	1.70688648	1.82351588	1.94780002	2.08022075	2.22128901
28	1.74102421	1.86454499	1.99649502	2.13742682	2.28792768
29	1.77584469	1.90649725	2.04640739	2.19620606	2.35656551
30	1.81136158	1.94939344	2.09756758	2.25660173	2.42726247
31	1.84758882	1.99325479	2.15000677	2.31865828	2.50008035
32	1.88454059	2.03810303	2.20375694	2.38242138	2.57508276
33	1.92223140	2.08396034	2.25885086	2.44793797	2.65233524
34	1.96067603	2.13084945	2.31532213	2.51525626	2.73190530
35	1.99988955	2.17879356	2.37320519	2.58442581	2.81386245
36	2.03988734	2.22781642	2.43253532	2.65549752	2.89827833
37	2.08068509	2.27794229	2.49334870	2.72852370	2.98522668
38	2.12229879	2.32919599	2.55568242	2.80355810	3.07478348
39	2.16474477	2.38160290	2.61957448	2.88065595	3.16702698
40	2.20803966	2.43518807	2.68506384	2.95987399	3.26203779
41	2.25220046	2.48998072	2.75219043	3.04127052	3.35989893
42	2.29724447	2.54600528	2.82099520	3.12490546	3.46069589
43	2.34318936	2.60329040	2.89152008	3.21084036	3.56451677
44	2.39005314	2.66186444	2.96380808	3.29913847	3.67145227
45	2.43785421	2.72175639	3.03790328	3.38986478	3.78159584
46	2.48661129	2.78299590	3.11385086	3.48308606	3.89504372
47	2.53634352	2.84561331	3.19169713	3.57887093	4.01189503
48	2.58707039	2.90963961	3.27148956	3.67728988	4.13225188
49	2.63881179	2.97510650	3.35327680	3.77841535	4.25621944
50	2.69158803	3.04204640	3.43710872	3.88232177	4.38390602

AMOUNT AT COMPOUND INTEREST $(1 + i)^n$
(Continued)

Years n	.02(2 %)	.0225(2¼ %)	.025(2½ %)	.0275(2¾ %)	.03(3 %)
	Rate i				
50	2.69158803	3.04204640	3.43710872	3.88232177	4.38390602
51	2.74541979	3.11049244	3.52303644	3.98908562	4.51542320
52	2.80032819	3.18047852	3.61111235	4.09878547	4.65088590
53	2.85633475	3.25203929	3.70139016	4.21150208	4.79041247
54	2.91346144	3.32521017	3.79392491	4.32731838	4.93412485
55	2.97173067	3.40002740	3.88877303	4.44631964	5.08214859
56	3.03116529	3.47652802	3.98599236	4.56859343	5.23461305
57	3.09178859	3.55474990	4.08564217	4.69422975	5.39165144
58	3.15362436	3.63473177	4.18778322	4.82332107	5.55340098
59	3.21669685	3.71651324	4.29247780	4.95596239	5.72000301
60	3.28103079	3.80013479	4.39978975	5.09225136	5.89160310
61	3.34665140	3.88563782	4.50978449	5.23228827	6.06835120
62	3.41358443	3.97306467	4.62252910	5.37617620	6.25040173
63	3.48185612	4.06245862	4.73809233	5.52402105	6.43791379
64	3.55149324	4.15386394	4.85654464	5.67593162	6.63105120
65	3.62252311	4.24732588	4.97795826	5.83201974	6.82998273
66	3.69497357	4.34289071	5.10240721	5.99240029	7.03488222
67	3.76887304	4.44060576	5.22996739	6.15719130	7.24592868
68	3.84425050	4.54051939	5.36071658	6.32651406	7.46330654
69	3.92113551	4.64268107	5.49473449	6.50049319	7.68720574
70	3.99955822	4.74714140	5.63210286	6.67925676	7.91782191
71	4.07954939	4.85395208	5.77290543	6.86293632	8.15535657
72	4.16114038	4.96316600	5.91722806	7.05166706	8.40001727
73	4.24436318	5.07483723	6.06515876	7.24558791	8.65201778
74	4.32925045	5.18902107	6.21678773	7.44484158	8.91157832
75	4.41583546	5.30577405	6.37220743	7.64957472	9.17892567
76	4.50415216	5.42515396	6.53151261	7.85993802	9.45429344
77	4.59423521	5.54721993	6.69480043	8.07608632	9.73792224
78	4.68611991	5.67203237	6.86217044	8.29817869	10.0300599
79	4.77984231	5.79965310	7.03372470	8.52637861	10.3309617
80	4.87543916	5.93014530	7.20956782	8.76085402	10.6408906
81	4.97294794	6.06357357	7.38980701	9.00177751	10.9601173
82	5.07240690	6.20000397	7.57455219	9.24932639	11.2889208
83	5.17385504	6.33950406	7.76391599	9.50368286	11.6275884
84	5.27733214	6.48214290	7.95801389	9.76503414	11.9764161
85	5.38287878	6.62799112	8.15696424	10.0335726	12.3357085
86	5.49053636	6.77712092	8.36088834	10.3094958	12.7057798
87	5.60034708	6.92960614	8.56991055	10.5930070	13.0869532
88	5.71235402	7.08552228	8.78415832	10.8843147	13.4795618
89	5.82660110	7.24494653	9.00376228	11.1836333	13.8839487
90	5.94313313	7.40795782	9.22885633	11.4911832	14.3004671
91	6.06199579	7.57463688	9.45957774	11.8071908	14.7294811
92	6.18323570	7.74506621	9.69606718	12.1318885	15.1713656
93	6.30690042	7.91933020	9.93846886	12.4655154	15.6265065
94	6.43303843	8.09751512	10.1869306	12.8083171	16.0953017
95	6.56169920	8.27970921	10.4416038	13.1605458	16.5781608
96	6.69293318	8.46600267	10.7026439	13.5224608	17.0755056
97	6.82679184	8.65648773	10.9702100	13.8943285	17.5877708
98	6.96332768	8.85125871	11.2444653	14.2764226	18.1154039
99	7.10259423	9.05041203	11.5255769	14.6690242	18.6588660
100	7.24464612	9.25404630	11.8137164	15.0724223	19.2186320

Reprinted with permission. From *Mathematical Tables from Handbook of Chemistry & Physics, Tenth Edition.* Copyright CRC Press, Boca Raton, Florida.

AMOUNT AT COMPOUND INTEREST $(1 + i)^n$

(Continued)

n	.035(3½ %)	.04(4 %)	.045(4½ %)	.05(5 %)	.055(5½ %)
			Rate i		
1	1.03500000	1.04000000	1.04500000	1.05000000	1.05500000
2	1.07122500	1.08160000	1.09202500	1.10250000	1.11302500
3	1.10871788	1.12486400	1.14116613	1.15762500	1.17424138
4	1.14752300	1.16985856	1.19251860	1.21550625	1.23882465
5	1.18768631	1.21665290	1.24618194	1.27628156	1.30696001
6	1.22925533	1.26531902	1.30226012	1.34009564	1.37884281
7	1.27227926	1.31593178	1.36086183	1.40710042	1.45467916
8	1.31680904	1.36856905	1.42210061	1.47745544	1.53468651
9	1.36289735	1.42331181	1.48609514	1.55132822	1.61909427
10	1.41059876	1.48024428	1.55296942	1.62889463	1.70814446
11	1.45996972	1.53945406	1.62285305	1.71033936	1.80209240
12	1.51106866	1.60103222	1.69588143	1.79585633	1.90120749
13	1.56395606	1.66507351	1.77219610	1.88564914	2.00577390
14	1.61869452	1.73167645	1.85194492	1.97993160	2.11609146
15	1.67534883	1.80094351	1.93528244	2.07892818	2.23247649
16	1.73398604	1.87298125	2.02237015	2.18287459	2.35526270
17	1.79467555	1.94790050	2.11337681	2.29201832	2.48480215
18	1.85748920	2.02581652	2.20847877	2.40661923	2.62146627
19	1.92250132	2.10684918	2.30786031	2.52695020	2.76564691
20	1.98978886	2.19112314	2.41171402	2.65329771	2.91775749
21	2.05943147	2.27876807	2.52024116	2.78596259	3.07823415
22	2.13151158	2.36991879	2.63365201	2.92526072	3.24753703
23	2.20611448	2.46471554	2.75216635	3.07152376	3.42615157
24	2.28332849	2.56330416	2.87601383	3.22509994	3.61458990
25	2.36324498	2.66583633	3.00543446	3.38635494	3.81339235
26	2.44595856	2.77246978	3.14067901	3.55567269	4.02312893
27	2.53156711	2.88336858	3.28200956	3.73345632	4.24440102
28	2.62017196	2.99870332	3.42969999	3.92012914	4.47784307
29	2.71187798	3.11865145	3.58403649	4.11613560	4.72412444
30	2.80679370	3.24339751	3.74531813	4.32194238	4.98395129
31	2.90503148	3.37313341	3.91385745	4.53803949	5.25806861
32	3.00670759	3.50805875	4.08998104	4.76494147	5.54726238
33	3.11194235	3.64838110	4.27403018	5.00318854	5.85236181
34	3.22086033	3.79431634	4.46636154	5.25334797	6.17424171
35	3.33359045	3.94608899	4.66734781	5.51601537	6.51382501
36	3.45026611	4.10393255	4.87737846	5.79181614	6.87208538
37	3.57102543	4.26808986	5.09686049	6.08140694	7.25005008
38	3.69601132	4.43881345	5.32621921	6.38547729	7.64880283
39	3.82537171	4.61636599	5.56589908	6.70475115	8.06948699
40	3.95925972	4.80102063	5.81636454	7.03998871	8.51330877
41	4.09783381	4.99306145	6.07810094	7.39198815	8.98154076
42	4.24125799	5.19278391	6.35161548	7.76158756	9.47552550
43	4.38970202	5.40049527	6.63743818	8.14966693	9.99667940
44	4.54334160	5.61651508	6.93612290	8.55715028	10.5464968
45	4.70235855	5.84117568	7.24824843	8.98500779	11.1265541
46	4.86694110	6.07482271	7.57441961	9.43425818	11.7385146
47	5.03728404	6.31781562	7.91526849	9.90597109	12.3841329
48	5.21358898	6.57052824	8.27145557	10.4012696	13.0652602
49	5.39606459	6.83334937	8.64367107	10.9213331	13.7838495
50	5.58492686	7.10668335	9.03263627	11.4673998	14.5419612

AMOUNT AT COMPOUND INTEREST $(1 + i)^n$
(Continued)

Years	Rate i				
n	.06(6 %)	.065(6½ %)	.07(7 %)	.075(7½) %	.08(8 %)
1	1.06000000	1.06500000	1.07000000	1.07500000	1.08000000
2	1.12360000	1.13422500	1.14490000	1.15562500	1.16640000
3	1.19101600	1.20794963	1.22504300	1.24229688	1.25971200
4	1.26247696	1.28646635	1.31079601	1.33546914	1.36048896
5	1.33822558	1.37008666	1.40255173	1.43562933	1.46932808
6	1.41851911	1.45914230	1.50073035	1.54330153	1.58687432
7	1.50363026	1.55398655	1.60578148	1.65904914	1.71382427
8	1.59384807	1.65499567	1.71818618	1.78347783	1.85093021
9	1.68947896	1.76257039	1.83845921	1.91723866	1.99900463
10	1.79084770	1.87713747	1.96715136	2.06103156	2.15892500
11	1.89829856	1.99915140	2.10485195	2.21560893	2.33163900
12	2.01219647	2.12909624	2.25219159	2.38177960	2.51817012
13	2.13292826	2.26748750	2.40984500	2.56041307	2.71962373
14	2.26090396	2.41487418	2.57853415	2.75244105	2.93719362
15	2.39655819	2.57184101	2.75903154	2.95887735	3.17216911
16	2.54035168	2.73901067	2.95216375	3.18079315	3.42594264
17	2.69277279	2.91704637	3.15881521	3.41935264	3.70001805
18	2.85433915	3.10665438	3.37993228	3.67580409	3.99601950
19	3.02559950	3.30858691	3.61652754	3.95148940	4.31570106
20	3.20713547	3.52364506	3.86968446	4.24785110	4.66095714
21	3.39956360	3.75268199	4.14056237	4.56643993	5.03383372
22	3.60353742	3.99660632	4.43040174	4.90892293	5.43654041
23	3.81974966	4.25638573	4.74052986	5.27709215	5.8714636
24	4.04893464	4.53305081	5.07236695	5.67287406	6.34118070
25	4.29187072	4.82769911	5.42743264	6.09833961	6.84847520
26	4.54938296	5.14149955	5.80735292	6.55571508	7.39635321
27	4.82234594	5.47569702	6.21386763	7.04739371	7.98806147
28	5.11168670	5.83161733	6.64883836	7.57594824	8.62710639
29	5.41838790	6.21067245	7.11425705	8.14414436	9.31727490
30	5.74349117	6.61436016	7.61225504	8.75495519	10.0626569
31	6.08810064	7.04429996	8.14511290	9.41157683	10.8676694
32	6.45338668	7.50217946	8.71527080	10.1174451	11.7370830
33	6.84058988	7.98982113	9.32533975	10.8762535	12.6760496
34	7.25102528	8.50915950	9.97811354	11.6919725	13.6901336
35	7.68608679	9.06225487	10 6765815	12.5688704	14.7853443
36	8.14725200	9.65130143	11.4239422	13.5115357	15.9681718
37	8.63608712	10.2786360	12.2236181	14.5249009	17.2456256
38	9.15425235	10.9467474	13.0792714	15.6142684	18.6252756
39	9.70350749	11.6582859	13.9948204	16.7853386	20.1152977
40	10.2857179	12.4160745	14.9744578	18.0442390	21.7245215
41	10.9028610	13.2231194	16.0226699	19.3975569	23.4624832
42	11.5570327	14.0826221	17.1442568	20.8523737	25.3394819
43	12.2504546	14.9979926	18.3443548	22.4163017	27.3666404
44	12.9854819	15.9728621	19.6284596	24.0975243	29.5559717
45	13.7646108	17.0110981	21.0024518	25.9048386	31 9204494
46	14.5904875	18.1168195	22.4726234	27.8477015	34 4740852
47	15.4659167	19.2944128	24.0457070	29.9362791	37 2320122
48	16.3938717	20.5485496	25.7289065	32.1815001	40.2105731
49	17.3775040	21.8842053	27.5299300	34.5951126	43.4274190
50	18.4201543	23.3066787	29.4570251	37.1897460	46.9016125

CHAPTER 4

Present Values

Chapter 3 showed how someone who started with a certain amount of money would answer the question, "How much will I have in a given time at a given interest rate?" This chapter covers the opposite case. Suppose you need a certain amount of money at a definite future time, and you have an interest rate you can earn. How much must you set aside now to make sure you have the required future amount at the required future time? This is called a present value. In this chapter, we look at the present value equations and at present value interest tables and we examine how the present values change as interest rate and time change.

We then use the present value concept to analyze a proposed project, so we can determine whether or not we should proceed with the project. We use this present value analysis of a project to develop the concept of internal rate of return. This is a common problem in many activities, including many business applications.

We will use the concept of present value throughout this book. It is the most important of the concepts we present in the chapters on compound interest functions, even more widely used than the compound interest concept. You should understand this chapter on present values thoroughly, including the present value equations, before you go on to the next part of the book.

This chapter also builds on the previous chapter. Mathematical developments presented in the previous chapter, such as compounding within a period and continuous compounding, will not be repeated here.

WHAT IS PRESENT VALUE?

Suppose you need a given amount of money at some given time in the future. You might want to know, "How much do I need to put aside now, at a given interest rate, to have the money I need at the time I need it?" Here are some examples.

You wish to provide a college education fund for your 2-year-old child. You figure you will need $100,000 in 16 years, and you can earn 6% interest, compounded yearly. How much should you put aside now?

You will need to buy a new car in 5 years. It will cost $20,000, and you can earn 5.5% while you wait. How much should you put aside now?

The amount you will need now to have a given future amount, at a given time, assuming an interest rate you can earn, is called the present value of the future amount.

THE EQUATION FOR PRESENT VALUE

Suppose you need $100 in one year, and the bank will pay you 6% interest, true annual rate, for the year. How much must you deposit now to have your $100 in one year? Let *PV* be the amount. We then have

$$PV \times (1+.06) = 100 \qquad \text{Equation 4.1}$$

Then

$$PV = 100/(1.06)$$
$$= 94.339623 \qquad \text{Equation 4.2}$$

We can check this answer:

$$(94.34)(1.06) = \$100.00$$

We can get the same result in a slightly different way. You start with 100 and end with 106. The present value factor will be 100/106, or .94339623.

THE GENERAL EQUATION FOR PRESENT VALUE

We will use the letter v to indicate a present value, so in the previous case, we have

$$v = 94.339623 \qquad\qquad \text{Equation 4.3}$$

or, for the present value of 1 in one year, at 6%, we have

$$v = .94339623 \qquad\qquad \text{Equation 4.4}$$

In general terms, we want to have v increase to 1 after t years. Then

$$v^t (1+i)^t = 1 \qquad\qquad \text{Equation 4.5}$$

and

$$v^t = \frac{1}{(1+i)^t} \qquad\qquad \text{Equation 4.6}$$

For n compoundings per year (or per period),

$$v^{Nt} = \frac{1}{\left(1+\dfrac{i}{n}\right)^{nt}} \qquad\qquad \text{Equation 4.7}$$

You can see that this equation is the reciprocal of the compound interest equation. Compounding, including compounding within a period, works exactly as in the compound interest equations in the previous chapter. Continuous functions also work exactly as they did in the previous chapter, and they would be derived in the same way. Here are two examples, solving the examples at the start of the chapter.

EXAMPLE 4.1.   In the case of the college fund, you will need $100,000 in 16 years, and you can earn 6% on your investment. You will need $100,000 $(1/1.06)^{16}$, or

$100,000 (.3936463), or
$39,364.63

Actually, you will probably want to make annual contributions to the fund. Most people, especially those with young children, don't have that much money. Such a series of payments is called an annuity certain. We'll cover annuities certain in the next chapter.   ∎

EXAMPLE 4.2.   In the case of the new car, you will need $20,000 in 5 years, and you can earn 5.5% on your investment. You will need $20,000 $(1/1.055)^5$, or

$20,000 (.765134), or
$15,302.68.   ■

THE PRESENT VALUE TABLES

Look at the present value tables on pages 68–75. You can see that as you move down any column, the values decrease. You have more time to earn compound interest, so you need to put aside less now to achieve the same amount. The present values approach zero, but do not actually ever reach it. Money always has a present value, no matter when it is due. For example, $1 due in 1 billion years has a present value, but not very much.

Similarly, as you move across a row, the values decrease. The interest rates increase, so you earn at a higher rate. You therefore need to put aside less money now to achieve the same final amount.

As in the tables for compound interest, you multiply your desired future amount by the present value factor for 1 to get the desired present value.

USING PRESENT VALUES TO MAKE PROJECT DECISIONS

Suppose you are a project manager, and you are presented with a project. How would you decide whether or not to proceed with the project?

You have a choice of two ways. One way is to take the present value of the outflows and the present value of the inflows, and compare the two values. If the present value of the inflows is greater than the present value of the out-flows, you proceed with the project.

A second way is to find the interest rate (or rates) that make(s) the present value of the outflows equal to the present value of the inflows. If the rate is high enough, you proceed with the project. This rate, which equalizes the present values of the inflows and the outflows, is called the internal rate of return (abbreviated IRR). The internal rate of return also makes the present value of the entire project equal to zero. It is widely used in project analysis. In finance, the yield of a bond is the internal rate of return of the bond payments and the bond purchase price.

Both methods require that you first decide on an interest rate that acts as a decision cutoff point. If you can earn more than this rate on the project, you proceed. If you cannot earn this rate, then you do not proceed. Of course, if your boss wants to do the project anyway, you proceed with the project.

EXAMPLE OF PROJECT ANALYSIS

Suppose you are given a project to examine. The project will cost $50,000 now and will yield a return according to the following schedule (amounts are at year-end for simplicity):

End of year	Amount
0 (now)	50,000 (–) (investment)
1	10,000
2	30,000
3	50,000
4	40,000
5	10,000

Assume that there is no salvage value after 5 years and no additional income or expense. Assume that your required return is 10% per year, true annual yield. We will disregard compounding within the year for simplicity.

THE FIRST METHOD

Find the present value factors for 10% for years 1 through 5, inclusive. Then figure the present values of the individual income stream receipts. Add up these present values, and see if the total is greater than $50,000.

Year	Present value factor	Present value of income
1	.909091	9,090.91
2	.826446	24,793.38
3	.751315	37,565.75
4	.683013	27,320.52
5	.620921	6,209.21
Total		104,979.77

The present value of the income is greater than the cost, at 10% per year, so you decide to proceed with the project.

THE SECOND METHOD

Find the rate that equalizes the present values of the outflows and the incomes. In this case, it is the rate that makes the present values of the inflows equal to 50,000. That rate is 43%. Here are the numbers:

Year	Cash flow	Present value factor	Present value of cash flow
0	50,000 (–)		
1	10,000	.69930	6,993.01
2	30,000	.48902	14,670.64
3	50,000	.34197	17,098.65
4	40,000	.23914	9,565.68
5	10,000	.16723	1,672.32
Total			50,000.30

(Totals may not add due to rounding.)

The IRR of 43% is higher than your required 10%, so you proceed with the project.

Here is another example, of the sort many people frequently encounter. You subscribe to a magazine and want to continue. The magazine offers a 1-year subscription for $32 and a 2-year subscription for $60. Assume that the subscription offer will not change for several years. Which subscription offer should you take?

Your investment now for the second year is $28 (= $60 for 2 years less $32 for the first year). In 1 year, this will be worth $32, the cost of a 1-year subscription. Then the present value of $32 in 1 year is $28. This works out to 14.29% per year.

$$\$28 \times 1.1429 = \$32$$

If this return is high enough for you, you take the 2-year subscription. Note that a more complete analysis would include the present value of the cost of postage, check, and envelope in 1 year. This would increase the rate of return a little.

Sometimes a flow of funds may have more than one solution which makes sense. Here is an example:

Year	Cash flow
0	–1,000
1	2,300
2	–1,320

This might represent a case where a machine must be changed to do a production run and then changed back to its original state when the production run is finished. The first expense, in year 0 (now) changes the machine. During the first year, a net gain of 2,300 is earned. During the second year, the machine must be changed back, with a net cost of 1,320. This might include some income, as well as the reset costs, combined.

This has two internal rates of return: 10% and 20%. Here are the figures:

Year	Amount	PV factor 10%	PV amount	PV factor 20%	PV amount
0	−1,000		−1,000.00		−1,000.00
1	2,300	.90909	2,090.91	.83333	1,916.67
2	−1,320	.82645	−1,090.91	.69444	− 916.67
Total			0.00		0.00

You can see that the internal rate of return can be either 10% or 20%; both IRRs are correct. You can also see that the example is of a sort that could easily occur in practical situations, and the rates of return are both well within the reasonable range.

In Chapter 3, we commented on the possibility of a negative interest rate. In the case of the internal rate of return, the IRR can easily be negative. This means that the project returns a loss. Here is an easy example:

End of year	Cash flow	
0 (now)	10 (−)	Investment
1	5	Return (all of it)

You can see that the internal rate of return is −.5, a loss of 50%. In this case, $i = -.5$, and $1/(1 + i) = 1/(1 - .5) = 1/(.5) = 2$. The present value of the return of 5 in 1 year is $5 \times 2 = 10$, the amount of the investment.

We'll discuss the whole question of solutions to this problem at greater length. It is important and deserves considerable attention.

USING DIFFERENT INTEREST RATES IN THE ANALYSIS

All the examples in this chapter, and the mathematical analyses, assume that all the fund flows are discounted at the same interest rate. Can you use different interest rates to compute the present values for some or all of the flows? Yes, you can. Later in the book, we'll show you how to develop interest rates you might use and how to analyze variable cash flows, uncertain cash flows, and use different interest rates to compute the present values. But if you use different rates, the concept of the internal rate of return no longer applies. IRR implies that you are using the same rate to evaluate all the present values. We'll cover this subject more completely in a later chapter.

THE EQUATIONS FOR FLOW OF FUNDS ANALYSIS

Suppose the years (or periods) are numbered sequentially, starting with number 0 (now), to n, where n is the number of the last year, or period. Let

a_j be the payment at time j, with sign + for inflows and sign − for outflows. Each payment at time j will be discounted back to the present using the factor $1/(1 + i)^j$, where i = the interest rate used, and j = the number of years (periods) until the payment is made.

The equation for internal rate of return is given by

$$\sum_{j=0}^{n} \frac{a_j}{(1+i)^j} = 0, \text{ for } j = 0 \text{ to } n \qquad\qquad \text{Equation 4.8}$$

This, in turn, can be stated as

$$\sum_{j=0}^{n} a_j y^j = 0, \text{ where } y = \frac{1}{(1+i)}, i \neq -1, j = 0 \text{ to } n \qquad \text{Equation 4.9}$$

We now have a polynomial of degree n, and we can apply the standard algebraic and numerical analysis methods of solving for the solutions to the polynomial. The fundamental theorem of algebra states that this equation will have n solutions, counting multiplicities, in the complex domain. (A solution to a polynomial equation is also called a "zero" of the equation.) This means that some or all of the solutions may be complex numbers. Multiplicity means that the same solution, if it occurs more than once, will be counted as many times as it occurs. For example, the equation

$$x^2 - 4x + 4 = 0 \qquad\qquad \text{Equation 4.10}$$

has solutions (zeros) +2 and +2, so the solution has multiplicity 2. The polynomial can be factored into $(x − 2)(x − 2) = 0$.

In order for this to hold, the coefficients a_j must all be real. This will be true in virtually all cases you are likely to encounter. We'll talk about the various number systems, including the real numbers, in the next section.

THE VARIOUS NUMBER SYSTEMS AND WHAT THEY MEAN

To understand the development of solutions to polynomial equations, you should understand the various kinds of number systems. However, many readers may not be familiar with the various number systems met in this study. This section will give you a quick view of the subject. It does not replace the many fine mathematics texts and does not offer rigorous proofs. For these, the reader should consult the various texts. This section is meant to give the reader an idea of what to look for, what the various terms mean, and some notion of how the various numbers work.

THE NATURAL NUMBERS

We first look at the numbers we learned to count with, the natural numbers. These are the positive integers, starting with 1, and continuing upward, thus 1, 2, 3, 4, ... There are an infinite number of natural numbers, but you can count them, in the sense that if the natural numbers are ordered in ascending sequence, one can start counting and eventually get to any given number, although it might take quite a while. The natural numbers are therefore said to be denumerably infinite.

THE INTEGERS

If we add 0 (zero) and the negatives of the natural numbers to the set of natural numbers, we get the integers. The integers look like this: . . . −3, −2, −1, 0, +1, +2, +3, ... There are an infinite number of integers. You can also see that the natural numbers are a subset of the integers. However, we can arrange the integers so that they can be counted, as follows: 0, +1, −1, +2, −2, ... You can see that the integers are also denumerably infinite, just like the natural numbers, because you can count them. This seems strange, but it is true, and was proved in the 19th century by German mathematician Georg Cantor. It came as a surprise to mathematicians that a subset of a set, which excludes some members of the original set, could equal the size of the original set. The size of a set is called its cardinality. The cardinality of both the natural numbers and the integers is said to be denumerably infinite. For example, a set with three members has cardinality 3; a set with 11 members has cardinality 11.

THE RATIONAL NUMBERS

If you divide one integer by another (except zero), you obtain a fraction. Any ratio of two integers can also be expressed in decimal form, either as a finite decimal (1/4 = .25) or as an infinite, repeating decimal (10/11 = .9090909 . . .). These numbers are called the rational numbers, and you can see that there are an infinite number of these also. Believe it or not, there are only a denumerably infinite number of rational numbers (Cantor proved this as well)—that is, there are just as many natural numbers as there are rational numbers. You can also see that the integers, as well as the natural numbers, are subsets of the rational numbers.

The rationals are the numbers that we use in everyday work, and usually not all of them, just the ones that can be represented by decimals that terminate soon after the decimal place. Most work requires a relatively small number of significant figures, and we terminate the numbers we use after we have enough precision. This is done either by rounding, the usual method, or by truncation, sometimes used in finance.

THE REAL NUMBERS

Many numbers are not rational numbers. For example, the square root of 2 cannot be expressed as the quotient of two integers (Euclid proved this). Numbers of this sort are called irrational. Some irrationals can be expressed algebraically, such as the square root of 2, or the cube root of 3. These are called algebraic numbers, and there are a denumerably infinite number of these as well. But other numbers, such as π (the relationship between the circumference of a circle and its diameter) and e (the base of natural logarithms) cannot be expressed even by algebraic means. You probably learned about π in the seventh grade and learned that it is approximately 3.14, although a better approximation is 3.1416. π has actually been calculated out to millions of decimal places. Numbers such as π and e are called transcendentals. The entire set containing both rational and irrational numbers is called the set of real numbers, and the system containing them is called the real number system. The reals can be matched with the points on a straight line, with each real number corresponding to a point on the line. The reals cannot be counted; they are nondenumerable, or nondenumerably infinite (also proved by Cantor).

Observe that even when we need a transcendental number for calculation purposes, we use a rational approximation for that number. For example, for π we might use 3.14, or 3.1416, or even carry it out much further, but we are still using rationals. We have to. The calculation must end sometime to be of any value, so we must use a rational approximation.

THE COMPLEX (OR IMAGINARY) NUMBERS

The reals don't contain a solution to many polynomial equations. In particular, they don't contain a solution to

$$x^2 + 1 = 0 \qquad\qquad \text{Equation 4.11}$$

To solve this equation, mathematicians invented the number i, which equals the square root of −1. Then i becomes the solution to Equation 4.11.

This led to the creation of what mathematicians call the complex number system. A complex number is a number in the form $a + bi$, where a and b are real numbers, and i is the square root of −1. Thus, for example, $1 + 2i$, $3 − 4i$,

1.7 + 3.79i, and 19i are all complex numbers. In particular, the real numbers can be thought of as the subset of complex numbers of the form a + 0i, where a is any real number. Note, however, that a + 0i is not the real number a; it is a complex number. In our example, a + bi, a is called the "real" part, and bi is called the "imaginary" part.

Be careful not to confuse the imaginary number i with the interest rate representation i used in this book. In this book, it will always be clear from the context which is meant.

Complex numbers can be thought of as coming in pairs, a + bi and a − bi. Such pairs are called complex conjugates.

You can see why the coefficients of the equation in the previous section will be real in almost all cases. For cash flow analysis, including project analysis, the coefficients represent cash inflows and outflows. These must be real; in fact, I cannot think of a case where they won't be rational. You probably don't want to tell your boss that the projected cash inflows (and outflows) are part real and part imaginary.

SOLVING POLYNOMIAL EQUATIONS

The equations we have been looking at have the general form as described earlier. The fundamental theorem of algebra assures us that any polynomial equation of degree n has n solutions, counting repetitions. However, it makes no statement about whether the solutions are real or complex, or about the signs of the solutions if they are real. If an equation has a complex solution $a + bi$, then the complex conjugate $a − bi$ is also a solution to the equation; that is, complex solutions occur in pairs, in the form of complex conjugates. This means that any polynomial whose highest exponent is odd must have at least one real solution. Of course, it could have more.

Generally, a solution will not interest the project analyst unless it is a positive real number. You probably don't want to tell your boss that the return is negative, or is part real and part imaginary.

Descartes's Rule of Signs gives the analyst a handy way of finding the number of positive zeros of a polynomial. The rule states that the number of positive zeros (n) is less than or equal to the number of variations (v) in sign of the coefficients of the polynomial. A variation of sign in a polynomial occurs when two terms next to each other have different signs. For example, in the polynomial $3x^3 − 4x^2 = 0$, there is a change of sign in the two terms with powers of x. In addition, the term $(v − n)$ is an even integer. Since $n \leq v$, $(v − n)$ cannot be negative, it must be greater or equal to zero. You find the number of sign variations by counting the number of sign changes in the nonzero coefficients of the polynomial.

This means that in the common case, where an investment (outflow) is followed only by inflows, $v = 1$. $v - n$ must be greater than zero, so that $n = 1$. This guaranties at least one positive solution. A bond investment, with the bond purchase (outflow) and receipts of interest and principal (inflows), is such a case. Many projects also have this feature. The investment in the project is expected to yield a positive flow of funds until the flow ends.

Thus (in this case) $y = 1/(1 + i)$ must be positive in at least one case. This does not mean that i must be positive. But it does guarantee that a real i will exist. It may be positive (you have made money), or negative (you have lost money), or zero (you have broken even), but it will exist. Remember, $i \neq -1$.

For bond investment, this means that if the total amount received (interest payments and maturity amount) is greater than the amount invested (the price plus accrued interest), there will be a positive return, or yield.

Here are two easy examples, one presented earlier in this chapter. They show two cases, where y is positive, but in one case i is negative, and in the other case i is positive.

Year	Cash flow	Present value factor	Present value of cash flow
0 (now)	10 (−)	1.00	−10
1	5	2.00	10
Total present value			0

You can see that the internal rate of return is −.5, a loss of 50%. In this case, $i = -.5$, and $y = 1/(1 + i) = 1/(1 - .5) = 1/(.5) = 2$. The present value of the return of 5 in one year is $5 \times 2 = 10$, the amount of the investment. Note that, the unknown in the equation, is positive, but i is negative.

Here is the second example:

Year	Cash flow	Present value factor	Present value of cash flow
0 (now)	10 (−)	1.00	−10
1	15	.6666667	10
Total present value			0

Here, y is still positive, but now i is also positive, in this case 50%. The project yields a 50% return after 1 year.

PRACTICAL CONSIDERATIONS IN USING CALCULATORS AND COMPUTERS TO SOLVE POLYNOMIAL EQUATIONS

Using computers or calculators to solve polynomial equations requires careful analysis. You cannot just "toss the problem into your computer," as one of my clients once suggested to me. The program may not produce solutions, even if solutions exist. For example, my calculator, one of the best and most popular, and on the market for more than 20 years, gave me "Error 3" when I tried to solve the three-payment problem presented earlier. The "Owner's Handbook and Problem Solving Guide" says that this indicates that the computation is very complex, may involve multiple answers, and cannot be continued until an estimate is provided. Yet, with just three terms, it would be difficult to produce a much simpler equation. The whole question of using computational methods to find answers belongs in the subject of numerical methods. These are mostly beyond the scope of this book. References are provided at the end of the chapter.

Many fine computer programs exist to help in these problems. These include Derive, Maple, MathCad, Mathematica, and MathKey.

Even so, the analyst should make sure to have all the solutions before continuing or showing the results to the boss. In most cases, this probably means finding all solutions, both real and complex, even if it takes a lot more work. The analyst should also check the computer's solutions by making sure that they actually solve the equation.

USING THE BISECTION METHOD TO FIND REAL SOLUTIONS

Usually, the analyst will only want the real solutions to the problem, and even those must be within some reasonable range. For example, if the polynomial has a solution of -10, the analyst probably won't be interested, at least in that solution. The analyst probably will only have an interest in positive real solutions.

One way of finding possible real solutions, if they exist, is to simply graph the function for all real values of x. The computer programs listed earlier will do this. If the graph crosses the x-axis, it has a real solution at the point of crossing. The bisection method, discussed previously, or other numerical analysis methods can then be used to find the actual numerical solution more precisely.

If you don't have a computer program, another way of finding positive real solutions, if they exist, is using the bisection method directly. Evaluate the polynomial for values along a range of hoped-for IRRs. For example, you might use values from .01 to .30, corresponding to interest rates 1% to 30%. Then check each pair of results to see whether they have different signs. If they do, there may be at least one solution between the two values. You can then use the bisection method to locate the solutions. Note that there could be two or more solutions between the values, which could result in no sign change, so further analysis is necessary. However, this method will frequently give some desired results.

Whatever method(s) you use, be sure you do a complete analysis of the polynomial to make sure you have correct results. You don't want a surprise when someone else produces different correct solutions than the ones you have presented.

WHAT IF THE EXPONENTS ARE NOT INTEGERS?

Suppose you have fractional exponents. This could happen if your inflows are not at even periods from each other, or from the outflows. The resulting equation is not a polynomial, so the previous analysis does not hold. What can you do?

1. You frequently can adjust the sizes and timing of the outflows and inflows to make them come annually. This is not as terrible as it sounds. Frequently, these amounts are subject to some estimation error anyway. Who can predict what a project's inflow might be 10 years from now? How can you predict when during a year the actual flow of funds might occur?

2. If the exponents of the individual terms differ by integral amounts, you might solve for the power of the unknown and then adjust the result. For example, if all the exponents are in the form $n + \frac{1}{2}$, substitute y^2 for x. The exponents will then all be integral, and you can solve for y. When you find all the solutions, take the square root(s) of each, and you have found the x's.

3. Try the bisection method. Plot the curve, using a computer program, and try solutions based on the estimated points where the curve crosses the x-axis.

CHAPTER SUMMARY

The present value of a future amount is the amount needed now to accumulate to the desired future amount, at a given interest rate and at the desired future time.

Present values can be used to analyze proposed projects, using one of two methods. Both methods require setting an interest rate to determine whether or not to proceed with the project.

One method is to evaluate inflows and outflows at the desired rate. If the present value of the inflows is greater than the present value of the outflows, proceed with the project.

A second method is to find the interest rate that makes the present values of the inflows and outflows equal. If this rate is higher than the desired rate, proceed with the project.

The rate that sets the present values of inflows and outflows equal is called the internal rate of return (IRR).

The inflows and outflows can be set up in equation form, which can frequently be solved.

Many such equations have more than one solution.

COMPUTER PROJECTS

1. What flows of funds, with outflow at time 0, and five inflows, at times 1 through 5, would give IRRs of 10%, 15%, and 20%? Plot them. How many different repetition possibilities exist? What additional IRRs could you add? Plot the curves, if possible.
2. Now make all five inflows the same. How many repetition possibilities exist? Plot them. What additional IRRs could you add? Plot the curves, if possible.

TOPICS FOR CLASS DISCUSSION

1. The Tappan Zee Bridge needs replacement. Several solutions have been proposed for a new bridge. One proposal (Proposal A) is to build a new bridge with an expected life of 50 years, similar to the present bridge. Another proposal (Proposal B) is to build a much improved bridge that will last much longer, perhaps forever. New York State Governor George L. Pataki has asked you to examine the problem. Assume that the Proposal A bridge will cost $500 million, that the Proposal B bridge (the "forever" bridge) will cost $ 1 billion, that maintenance costs of the bridges would be the same, and that the capital costs 6% per year. Which proposal do you recommend and why? What other factors would you need to consider, especially in your analysis of the "forever" bridge?
2. New York State Comptroller Alan G. Hevesi has told you that New York State can borrow all it wants at 4%. How does this affect your analysis in Problem 1?
3. A group of citizens is recommending a tunnel under the Hudson River to replace the bridge. They claim that a tunnel could carry rail lines, as well as highways, to connect to the MetroNorth Hudson Line for easy rail transport to New York City. The tunnel would be less subject

to delay due to bad weather. A tunnel might also have lower maintenance. What factors would you consider in your analysis of this project for the governor and the comptroller?

4. A leading business magazine has the following subscription rates:

1 year	$54.95
2 years	79.95
3 years	99.95

You want to subscribe to the magazine for several years. Which subscription should you use and why?

PROBLEMS

1. You wish to buy a new car in 4 years. You estimate the car will cost $17,000. A local bank will pay you 5.25%, compounded quarterly, on a 4-year certificate of deposit (CD). How much should you put into the CD to have enough to buy the car?

2. The dealer says that he expects the price of the car to increase 3% per year (true annual rate) over the next 4 years. The local bank will still pay you 5.25%, compounded quarterly, on a 4-year CD. Now how much should you put into the CD to have enough to buy the car?

3. Your assistant has proposed a project for your department to implement. The project requires an investment of $75,000, with inflows as in the following table:

Year	Cash flow
0 (now)	−75,000 (investment)
1	10,000
2	25,000
3	50,000
4	50,000
5	35,000

Corporate guidelines require an IRR of at least 15% true annual return for any project to be implemented. Assume that after 5 years, there are no further inflows or outflows for the project. Should you do the project?

4. Your assistant has come in and sheepishly told you that he made a mistake on the last three amounts. They should each be 30,000, instead of the numbers 50,000, 50,000, and 35,000, which he originally gave you. Should you still do the project?

5. What is the equation for present value?

SUGGESTIONS FOR FURTHER STUDY

For complex numbers, the following is a well-known and widely used text: Ahlfors, Lars V., *Complex Analysis*, 3rd ed., McGraw-Hill, New York, 1979.

A newer text, with a somewhat different approach, is Needham, Tristan, *Visual Complex Analysis*, Oxford University Press, New York, 1997.

A standard and widely used text on mathematical analysis is Rudin, Walter, *Principles of Mathematical Analysis*, 3rd ed., McGraw-Hill Inc., New York, 1976.

The question of whether mathematics is invented or discovered belongs in the philosophy of mathematics, and is apparently quite controversial. I believe that mathematics is a human construction, like skyscrapers, automobiles, music, the telephone, TV, computers, language, and many other inventions. Discussion of this includes the existence of the Platonic ideal and other considerations, and is far beyond the scope of this book. For further reading, see Hersh, Reuben, *What Is Mathematics, Really?*, Oxford University Press, New York, 1997.

PRESENT VALUE $1/(1 + i)^n$

The following table gives the value of unit amount due in n years at rate of interest i, compounded annually, $1/(1 + i)^n = v^n$.

Years n	.0025($\frac{1}{4}$ %)	.004167($\frac{1}{12}$ %)	.005($\frac{1}{2}$ %)	.005833($\frac{7}{12}$ %)	.0075($\frac{3}{4}$ %)
1	.99750623	.99585062	99502488	.99420050	.99255583
2	.99501869	.99171846	.99007450	.98843463	.98516708
3	.99253734	.98760345	.98514876	.98270220	.97783333
4	.99006219	.98350551	.98024752	.97700301	.97055417
5	.98759321	.97942457	.97537067	.97133688	.96332920
6	.98513038	.97536057	.97051808	.96570361	.95615802
7	.98267370	.97131343	.96568963	.96010301	.94904022
8	.98022314	.96728308	.96088520	.95453489	.94197540
9	.97777869	.96326946	.95610468	.94899906	.93496318
10	.97534034	.95927249	.95134794	.94349534	.92800315
11	.97290807	.95529211	.94661487	.93802354	.92109494
12	.97048187	.95132824	.94190534	.93258347	.91423815
13	.96806171	.94738082	.93721924	.92717495	.90743241
14	.96564759	.94344978	.93255646	.92179779	.90067733
15	.96323949	.93953505	.92791688	.91645182	.89397254
16	.96083740	.93563657	.92330037	.91113686	.88731766
17	.95844130	.93175426	.91870684	.90585272	.88071231
18	.95605117	.92788806	.91413616	.90059922	.87415614
19	.95366700	.92403790	.90958822	.89537619	.86764878
20	.95128878	.92020372	.90506290	.89018346	.86118985
21	.94891649	91638544	.90056010	.88502084	.85477901
22	.94655011	.91258301	.89607971	.87988815	.84841589
23	.94418964	.90879636	.89162160	.87478524	.84210014
24	.94183505	.90502542	.88718567	.86971192	.83583140
25	.93948634	.90127013	.88277181	.86466802	.82960933
26	.93714348	.89753042	.87837991	.85965338	.82343358
27	.93480646	.89380623	.87400986	.85466782	.81730380
28	.93247527	.89009749	.86966155	.84971117	.81121966
29	.93014990	.88640414	.86533488	.84478327	.80518080
30	.92783032	.88272611	.86102973	.83988394	.79918690
31	.92551653	.87906335	.85674600	.83501303	.79323762
32	.92320851	.87541578	.85248358	.83017037	.78733262
33	.92090624	.87178335	.84824237	.82535580	.78147158
34	.91860972	.86816599	.84402226	.82056914	.77565418
35	.91631892	.86456365	.83982314	.81581025	.76988008
36	.91403384	.86097624	.83564492	.81107896	.76414896
37	.91175445	.85740373	.83148748	.80637510	.75846051
38	.90948075	.85384604	.82735073	.80169853	.75281440
39	.90721272	.85030311	.82323455	.79704907	.74721032
40	.90495034	.84677488	.81913886	.79242659	.74164796
41	.90269361	.84326129	.81506354	.78783091	.73612701
42	.90044250	.83976228	.81100850	.78326188	.73064716
43	.89819701	.83627779	.80697363	.77871935	.72520809
44	.89595712	.83280776	.80295884	.77420316	.71980952
45	.89372281	.82935212	.79896402	.76971317	.71445114
46	.89149407	.82591083	.79498907	.76524922	.70913264
47	.88927090	.82248381	.79103390	.76081115	.70385374
48	.88705326	.81907102	.78709841	.75639883	.69861414
49	.88484116	81567238	.78318250	.75201209	.69341353
50	.88263457	.81228785	.77928607	.74765079	.68825165

PRESENT VALUE $1/(1 + i)^n$ (Continued)

Years n	.0025($\frac{1}{4}$ %)	.004167($\frac{1}{12}$ %)	.005($\frac{1}{2}$ %)	.005833($\frac{7}{12}$ %)	.0075($\frac{3}{4}$ %)
50	.88263457	.81228785	.77928607	.74765079	.68825165
51	.88043349	.80891736	.77540902	.74331479	.68312819
52	.87823790	.80556086	.77155127	.73900393	.67804286
53	.87604778	.80221828	.76771270	.73471808	.67299540
54	.87386312	.79888957	.76389324	.73045708	.66798551
55	.87168391	.79557468	.76009277	.72622079	.66301291
56	.86951013	.79227354	.75631122	.72200907	.65807733
57	.86734178	.78898610	.75254847	.71782178	.65317849
58	.86517883	.78571230	.74880445	.71365877	.64831612
59	.86302128	.78245208	.74507906	.70951990	.64348995
60	.86086911	.77920539	.74137220	.70540504	.63869970
61	.85872230	.77597217	.73768378	.70131404	.63394511
62	.85658085	.77275237	.73401371	.69724677	.62922592
63	.85444474	.76954593	.73036190	.69320308	.62454185
64	.85231395	.76635279	.72672826	.68918285	.61989266
65	.85018848	.76317291	.72311269	.68518593	.61527807
66	.84806831	.76000621	.71951512	.68121219	.61069784
67	.84595343	.75685266	.71593544	.67726150	.60615170
68	.84384382	.75371219	.71237357	.67333372	.60163940
69	.84173947	.75058476	.70882943	.66942872	.59716070
70	.83964037	.74747030	.70530291	.66554637	.59271533
71	.83754650	.74436876	.70179394	.66168653	.58830306
72	.83545786	.74128009	.69830243	.65784908	.58392363
73	.83337442	.73820424	.69482829	.65403388	.57957681
74	.83129618	.73514115	.69137143	.65024081	.57526234
75	.82922312	.73209078	.68793177	.64646973	.57097999
76	.82715523	.72905306	.68450923	.64272053	.56672952
77	.82509250	.72602794	.68110371	.63899306	.56251069
78	.82303491	.72301537	.67771513	.63528723	.55832326
79	.82098246	.72001531	.67434342	.63160288	.55416701
80	.81893512	.71702770	.67098847	.62793989	.55004170
81	.81689289	.71405248	.66765022	.62429816	.54594710
82	.81485575	.71108960	.66432858	.62067754	.54188297
83	.81282369	.70813902	.66102346	.61707792	.53784911
84	.81079670	.70520069	.65773479	.61349917	.53384527
85	.80877476	.70227454	.65446248	.60994118	.52987123
86	.80675787	.69936054	.65120644	.60640382	.52592678
87	.80474600	.69645863	.64796661	.60288698	.52201169
88	.80273915	.69356876	.64474290	.59939054	.51812575
89	.80073731	.69069088	.64153522	.59591437	.51426873
90	.79874046	.68782495	.63834350	.59245836	.51044043
91	.79674859	.68497090	.63516766	.58902240	.50664063
92	.79476168	.68212870	.63200763	.58560636	.50286911
93	.79277973	.67929829	.62886331	.58221014	.49912567
94	.79080273	.67647962	.62573464	.57883361	.49541009
95	.78883065	.67367265	.62262153	.57547666	.49172217
96	.78686349	.67087733	.61952391	.57213918	.48806171
97	.78490124	.66809361	.61644170	.56882106	.48442850
98	.78294388	.66532143	.61337483	.56552218	.48082233
99	.78099140	.66256076	.61032321	.56224243	.47724301
100	.77904379	..65981155	.60728678	.55898171	.47369033

Reprinted with permission. From *Mathematical Tables from Handbook of Chemistry & Physics, Tenth Edition.* Copyright CRC Press, Boca Raton, Florida.

PRESENT VALUE $1/(1 + i)^n$ (Continued)

Years n	.01(1 %)	.01125(1⅛ %)	.0125(1¼ %)	.015(1½ %)	.0175(1¾ %)
1	.99009901	.98887515	.98765432	.98522167	.98280098
2	.98029605	.97787407	.97546106	.97066175	.96589777
3	.97059015	.96699537	.96341833	.95631699	.94928528
4	.96098034	.95623770	.95152428	.94218423	.93295851
5	.95146569	.94559970	.93977706	.92826033	.91691254
6	.94204524	.93508005	.92817488	.91454219	.90114254
7	.93271805	.92467743	.91671593	.90102679	.88564378
8	.92348322	.91439054	.90539845	.88771112	.87041157
9	.91433982	.90421808	.89422069	.87459224	.85544135
10	.90528695	.89415880	.88318093	.86166723	.84072860
11	.89632372	.88421142	.87227746	.84893323	.82626889
12	.88744923	.87437470	.86150860	.83638742	.81205788
13	.87866260	.86464742	.85087269	.82402702	.79809128
14	.86996297	.85502835	.84036809	.81184928	.78436490
15	.86134947	.84551629	.82999318	.79985150	.77087459
16	.85282126	.83611005	.81974635	.78803104	.75761631
17	.84437749	.82680846	.80962602	.77638526	.74458605
18	.83601731	.81761034	.79963064	.76491159	.73177990
19	.82773992	.80851455	.78975866	.75360747	.71919401
20	.81954447	.79951995	.78000855	.74247042	.70682458
21	.81143017	.79062542	.77037881	.73149795	.69466789
22	.80339621	.78182983	.76086796	.72068763	.68272028
23	.79544179	.77313210	.75147453	.71003708	.67097817
24	.78756613	.76453112	.74219707	.69954392	.65943800
25	.77976844	.75602583	.73303414	.68920583	.64809632
26	.77204796	.74761516	.72398434	.67902052	.63694970
27	.76440392	.73929806	.71504626	.66898574	.62599479
28	.75683557	.73107348	.70621853	.65909925	.61522829
29	.74934215	.72294040	.69749978	.64935887	.60464697
30	.74192292	.71489780	.68888867	.63976243	.59424764
31	.73457715	.70694467	.68038387	.63030781	.58402716
32	.72730411	.69908002	.67198407	.62099292	.57398247
33	.72010307	.69130287	.66368797	.61181568	.56411053
34	.71297334	.68361223	.65549429	.60277407	.55440839
35	.70591420	.67600715	.64740177	.59386608	.54487311
36	.69892495	.66848667	.63940916	.58508974	.53550183
37	.69200490	.66104986	.63151522	.57644309	.52629172
38	.68515337	.65369578	.62371873	.56792423	.51724002
39	.67836967	.64642352	.61601850	.55953126	.50834400
40	.67165314	.63923216	.60841334	.55126232	.49960098
41	.66500311	.63212080	.60090206	.54311559	.49100834
42	.65841892	.62508855	.59348352	.53508925	.48256348
43	.65189992	.61813454	.58615656	.52718153	.47426386
44	.64544546	.61125789	.57892006	.51939067	.46610699
45	.63905492	.60445774	.57177290	.51171494	.45809040
46	.63272764	.59773324	.56471397	.50415265	.45021170
47	.62646301	.59108355	.55774219	.49670212	.44246850
48	.62026041	.58450784	.55085649	.48936170	.43485848
49	.61411921	.57800528	.54405579	.48212975	.42737934
50	.60803882	.57157506	.53733905	.47500468	.42002883

PRESENT VALUE $1/(1 + i)^n$ (Continued)

Years n	.01(1 %)	.01125(1⅛ %)	.0125(1¼ %)	.015(1½ %)	.0175(1¾ %)
			Rate i		
50	.60803882	.57157506	.53733905	.47500468	.42002883
51	.60201864	.56521637	.53070524	.46798491	.41280475
52	.59605806	.55892843	.52415332	.46106887	.40570492
53	.59015649	.55271044	.51768229	.45425505	.39872719
54	.58431336	.54656162	.51129115	.44754192	.39186947
55	.57852808	.54048120	.50497892	.44092800	.38512970
56	.57280008	.53446843	.49874461	.43441182	.37850585
57	.56712879	.52852256	.49258727	.42799194	.37199592
58	.56151365	.52264282	.48650594	.42166694	.36559796
59	.55595411	.51682850	.48049970	.41543541	.35931003
60	.55044962	.51107887	.47456760	.40929597	.35313025
61	.54499962	.50539319	.46870874	.40324726	.34705676
62	.53960054	.49977077	.46292222	.39728794	.34108772
63	.53426097	.49421090	.45720713	.39141669	.33522135
64	.52897126	.48871288	.45156259	.38563221	.32945587
65	.52373392	.48327602	.44598775	.37993321	.32378956
66	.51854844	.47789965	.44048173	.37431843	.31822069
67	.51341429	.47258309	.43504368	.36878663	.31274761
68	.50833099	.46732568	.42967277	.36333658	.30736866
69	.50329801	.46212675	.42436817	.35796708	.30208222
70	.49831486	.45698566	.41912905	.35267692	.29688670
71	.49338105	.45190177	.41395462	.34746495	.29178054
72	.48849609	.44687443	.40884407	.34233000	.28676221
73	.48365949	.44190302	.40379661	.33727093	.28183018
74	.47887078	.43698692	.39881147	.33228663	.27698298
75	.47412949	.43212551	.39388787	.32737599	.27221914
76	.46943514	.42731818	.38902506	.32253793	.26753724
77	.46478726	.42256433	.38422228	.31777136	.26293586
78	.46018541	.41786337	.37947879	.31307523	.25841362
79	.45562912	.41321470	.37479387	.30844850	.25396916
80	.45111794	.40861775	.37016679	.30389015	.24960114
81	.44665142	.40407194	.36559683	.29939916	.24530825
82	.44222913	.39957670	.36108329	.29497454	.24108919
83	.43785063	.39513148	.35662547	.29061531	.23694269
84	.43351547	.39073570	.35222268	.28632050	.23286751
85	.42922324	.38638882	.34787426	.28208917	.22886242
86	.42497350	.38209031	.34357951	.27792036	.22492621
87	.42076585	.37783961	.33933779	.27381316	.22105770
88	.41659985	.37363621	.33514843	.26976666	.21725572
89	.41247510	.36947956	.33101080	.26577996	.21351914
90	.40839119	.36536916	.32692425	.26185218	.20984682
91	.40434771	.36130448	.32288814	.25798245	.20623766
92	.40034427	.35728503	.31890187	.25416990	.20269057
93	.39638046	.35331029	.31496481	.25041369	.19920450
94	.39245590	.34937976	.31107636	.24671300	.19577837
95	.38857020	.34549297	.30723591	.24306699	.19241118
96	.38472297	.34164941	.30344287	.23947487	.18910190
97	.38091383	.33784861	.29969666	.23593583	.18584953
98	.37714241	.33409010	.29599670	.23244909	.18265310
99	.37340832	.33037340	.29234242	.22901389	.17951165
100	.36971121	.32669805	.28873326	.22562944	.17642422

PRESENT VALUE $1/(1 + i)^n$ (Continued)

Years n	$.02(2\%)$	$.0225(2\frac{1}{4}\%)$	$.025(2\frac{1}{2}\%)$	$.0275(2\frac{3}{4}\%)$	$.03(3\%)$
			Rate i		
1	.98039216	.97799511	.97560976	.97323601	.97087379
2	.96116878	.95647444	.95181440	.94718833	.94259591
3	.94232233	.93542732	.92859941	.92183779	.91514166
4	.92384543	.91484335	.90595064	.89716573	.88848705
5	.90573081	.89471232	.88385429	.87315400	.86260878
6	.88797138	.87502427	.86229687	.84978491	.83748426
7	.87056018	.85576946	.84126524	.82704128	.81309151
8	.85349037	.83693835	.82074657	.80490635	.78940923
9	.83675527	.81852161	.80072836	.78336385	76641673
10	.82034830	.80051013	.78119840	.76239791	.74409391
11	.80426304	.78289499	.76214478	.74199310	.72242128
12	.78849318	.76566748	.74355589	.72213440	.70137988
13	.77303253	.74881905	.72542038	.70280720	.68095134
14	.75787502	.73234137	.70772720	.68399728	.66111781
15	.74301473	.71622628	.69046556	.66569078	.64186195
16	.72844581	.70046580	.67362493	.64787424	.62316694
17	.71416256	.68505212	.65719506	.63053454	.60501645
18	.70015937	.66997763	.64116591	.61365892	.58739461
19	.68643076	.65523484	.62552772	.59723496	.57028603
20	.67297133	.64081647	.61027094	.58125057	.55367575
21	.65977582	.62671538	.59538629	.56569398	.53754928
22	.64683904	.61292457	.58086647	.55055375	.52189250
23	.63415592	.59943724	.56669724	.53581874	.50669175
24	.62172149	.58624668	.55287535	.52147809	.49193374
25	.60953087	.57334639	.53939059	.50752126	.47760557
26	.59757928	.56072997	.52623472	.49393796	.46369473
27	.58586204	.54839117	.51339973	.48071821	.45018906
28	.57437455	.53632388	.50087778	.46785227	.43707675
29	.56311231	.52452213	.48866125	.45533068	.42434636
30	.55207089	.51298008	.47674269	.44314421	.41198676
31	.54124597	.50169201	.46511481	.43128391	.39998715
32	.53063330	.49065233	.45377055	.41974103	.38833703
33	.52022873	.47985558	.44270298	.40850708	.37702625
34	.51002817	.46929641	.43190534	.39757380	.36604490
35	.50002761	.45896960	.42137107	.38693314	.35538340
36	.49022315	.44887002	.41109372	.37657727	.34503243
37	.48061093	.43899268	.40106705	.36649856	.33498294
38	.47118719	.42933270	.39128492	.35668959	.32522615
39	.46194822	.41988528	.38174139	.34714316	.31575355
40	.45289042	.41064575	.37243062	.33785222	.30655684
41	.44401021	.40160954	.36334695	.32880995	.29762800
42	.43530413	.39277216	.35448483	.32000968	.28895922
43	.42676875	.38412925	.34583886	.31144495	.28054294
44	.41840074	.37567653	.33740376	.30310944	.27237178
45	.41019680	.36740981	.32917440	.29499702	.26443862
46	.40215373	.35932500	.32114576	.28710172	.25673653
47	.39426836	.35141809	.31331294	.27941773	.24925876
48	.38653761	.34368518	.30567116	.27193940	.24199880
49	.37895844	.33612242	.29821576	.26466122	.23495029
50	.37152788	.32872608	.29094221	.25757783	.22810708

PRESENT VALUE $1/(1 + i)^n$ (Continued)

Years					
			Rate i		
n	.02(2 %)	.0225(2¼ %)	.025(2½ %)	.0275(2¾ %)	.03(3 %)
50	.37152788	.32872608	.29094221	.25757783	.22810708
51	.36424302	.32149250	.28384606	.25068402	.22146318
52	.35710100	.31441810	.27692298	.24397471	.21501280
53	.35009902	.30749936	.27016876	.23744497	.20875029
54	.34323433	.30073287	.26357928	.23109000	.20267019
55	.33650425	.29411528	.25715052	.22490511	.19676717
56	.32990613	.28764330	.25087855	.21888575	.19103609
57	.32343738	.28131374	.24475956	.21302749	.18547193
58	.31709547	.27512347	.23878982	.20732603	.18006984
59	.31087791	.26906940	.23296568	.20177716	.17482508
60	.30478227	.26314856	.22728359	.19637679	.16973309
61	.29880614	.25735801	.22174009	.19112097	.16478941
62	.29294720	.25169487	.21633179	.18600581	.15998972
63	.28720314	.24615635	.21105541	.18102755	.15532982
64	.28157170	.24073971	.20590771	.17618253	.15080565
65	.27605069	.23544226	.20088557	.17146718	.14641325
66	.27063793	.23026138	.19598593	.16687804	.14214879
67	.26533130	.22519450	.19120578	.16241172	.13800853
68	.26012873	.22023912	.18654223	.15806493	.13398887
69	.25502817	.21539278	.18199241	.15383448	.13008628
70	.25002761	.21065309	.17755358	.14971726	.12629736
71	.24512511	.20601769	.17322300	.14571023	.12261880
72	.24031874	.20148429	.16899805	.14181044	.11904737
73	.23560661	.19705065	.16487615	.13801503	.11557998
74	.23098687	.19271458	.16085478	.13432119	.11221357
75	.22645771	.18847391	.15693149	.13072622	.10894521
76	.22201737	.18432657	.15310389	.12722747	.10577205
77	.21766408	.18027048	.14936965	.12382235	.10269131
78	.21339616	.17630365	.14572649	.12050837	.09970030
79	.20921192	.17242411	.14217218	.11728309	.09679641
80	.20510973	.16862993	.13870457	.11414412	.09397710
81	.20108797	.16491925	.13532153	.11108917	.09123990
82	.19714507	.16129022	.13202101	.10811598	.08858243
83	.19327948	.15774105	.12880098	.10522237	.08600236
84	.18948968	.15426997	.12565949	.10240620	.08349743
85	.18577420	.15087528	.12259463	.09966540	.08106547
86	.18213157	.14755528	.11960452	.09699795	.07870434
87	.17856036	.14430835	.11668733	.09440190	.07641198
88	.17505018	.14113286	.11384130	.09187533	.07418639
89	.17162665	.13802724	.11106468	.08941638	.07202562
90	.16826142	.13498997	.10835579	.08702324	.06992779
91	.16496217	.13201953	.10571296	.08469415	.06789105
92	.16172762	.12911445	.10313460	.08242740	.06591364
93	.15855649	.12627331	.10061912	.08022131	.06399383
94	.15544754	.12349468	.09816500	.07807427	.06212993
95	.15239955	.12077719	.09577073	.07598469	.06032032
96	.14941132	.11811950	.09343486	.07395104	.05856342
97	.14648169	.11552029	.09115596	.07197181	.05685769
98	.14360950	.11297828	.08893264	.07004556	.05520164
99	.14079363	.11049221	.08676355	.06817086	.05359383
100	.13803297	.10806084	.08464737	.06634634	.05203284

PRESENT VALUE $1/(1 + i)^n$ (Continued)

Years n	.035(3½ %)	.04(4 %)	.045(4½ %)	.05(5 %)	.055(5½ %)
			Rate i		
1	.96618357	.96153846	.95693780	.95238095	.94786730
2	.93351070	.92455621	.91572995	.90702948	.89845242
3	.90194271	.88899636	.87629660	.86383760	.85161366
4	.87144223	.85480419	.83856134	.82270247	.80721674
5	.84197317	.82192711	.80245105	.78352617	.76513435
6	.81350064	.79031453	.76789574	.74621540	.72524583
7	.78599096	.75991781	.73482846	.71068133	.68743681
8	.75941156	.73069021	.70318513	.67683936	.65159887
9	.73373097	.70258674	.67290443	.64460892	.61762926
10	.70891881	.67556417	.64392768	.61391325	.58543058
11	.68494571	.64958093	.61619874	.58467929	.55491050
12	.66178330	.62459705	.58966386	.55683742	.52598152
13	.63940415	.60057409	.56427164	.53032135	.49856068
14	.61778179	.57747508	.53997286	.50506795	.47256937
15	.59689062	.55526450	.51672044	.48101710	.44793305
16	.57670591	.53390818	.49446932	.45811152	.42458109
17	.55720378	.51337325	.47317639	.43629669	.40244653
18	.53836114	.49362812	.45280037	.41552065	.38146590
19	.52015569	.47464242	.43330179	.39573396	.36157906
20	.50256588	.45638695	.41464286	.37688948	.34272896
21	.48557090	.43883360	.39678743	.35894236	.32486158
22	.46915063	.42195539	.37970089	.34184987	.30792567
23	.45328563	.40572633	.36335013	.32557131	.29187267
24	.43795713	.39012147	.34770347	.31006791	.27665656
25	.42314699	.37511680	.33273060	.29530277	.26223370
26	.40883767	.36068923	.31840248	.28124073	.24856275
27	.39501224	.34681657	.30469137	.26784832	.23560450
28	.38165434	.33347747	.29157069	.25509364	.22332181
29	.36874815	.32065141	.27901502	.24294632	.21167944
30	.35627841	.30831867	.26700002	.23137745	.20064402
31	.34423035	.29646026	.25550241	.22035947	.19018390
32	.33258971	.28505794	.24449991	.20986617	.18026910
33	.32134271	.27409417	.23397121	.19987254	.17087119
34	.31047605	.26355209	.22389589	.19035480	.16196321
35	.29997686	.25341547	.21425444	.18129029	.15351963
36	.28983272	.24366872	.20502817	.17265741	.14551624
37	.28003161	.23429685	.19619921	.16443563	.13793008
38	.27056194	.22528543	.18775044	.15660536	.13073941
39	.26141250	.21662061	.17966549	.14914797	.12392362
40	.25257247	.20828904	.17192870	.14204568	.11746314
41	.24403137	.20027793	.16452507	.13528160	.11133947
42	.23577910	.19257493	.15744026	.12883962	.10553504
43	.22780590	.18516820	.15066054	.12270440	.10003322
44	.22010231	.17804635	.14417276	.11636133	.09481822
45	.21265924	.17119841	.13796437	.11129651	.08987509
46	.20546787	.16461386	.13202332	.10599668	.08518965
47	.19851968	.15828256	.12633810	.10094921	.08074849
48	.19180645	.15219476	.12089771	.09614211	.07653885
49	.18532024	.14634112	.11569158	.09156391	.07254867
50	.17905337	.14071262	.11070965	.08720373	.06876652

Reprinted with permission. From *Mathematical Tables from Handbook of Chemistry & Physics, Tenth Edition.* Copyright CRC Press, Boca Raton, Florida.

PRESENT VALUE $1/(1 + i)^n$ (Continued)

Years n	.06 (6 %)	.065 (6½ %)	.07 (7 %)	.075 (7½ %)	.08 (8 %)
			Rate i		
1	.94339623	.93896714	.93457944	.93023256	.92592593
2	.88999644	.88165928	.87343873	.86533261	.85733882
3	.83961928	.82784909	.81629788	.80496057	.79383224
4	.79209366	.77732309	.76289521	.74880053	.73502985
5	.74725817	.72988084	.71298618	.69655863	.68058320
6	.70496054	.68533412	.66634222	.64796152	.63016963
7	.66505711	.64350621	.62274974	.60275490	.58349040
8	.62741237	.60423119	.58200910	.56070223	.54026888
9	.59189846	.56735323	.54393374	.52158347	.50024897
10	.55839478	.53272604	.50834929	.48519393	.46319349
11	.52678753	.50021224	.47509280	.45134319	.42888286
12	.49696936	.46968285	.44401196	.41985413	.39711376
13	.46883902	.44101676	.41496445	.39056198	.36769792
14	.44230096	.41410025	.38781724	.36331347	.34046104
15	.41726506	.38882652	.36244602	.33796602	.31524170
16	.39364628	.36509533	.33873460	.31438699	.29189047
17	.37136442	.34281251	.31657439	.29245302	.27026895
18	.35034379	.32188969	.29586392	.27204932	.25024903
19	.33051301	.30224384	.27650833	.25306913	.23171206
20	.31180473	.28379703	.25841900	.23541315	.21454821
21	.29415540	.26647608	.24151309	.21898897	.19865575
22	.27750510	.25021228	.22571317	.20371067	.18394051
23	.26179726	.23494111	.21094688	.18949830	.17031528
24	.24697855	.22060198	.19714662	.17627749	.15769934
25	.23299863	.20713801	.18424918	.16397906	.14601790
26	.21981003	.19449579	.17219549	.15253866	.13520176
27	.20736795	.18262515	.16093037	.14189643	.12518682
28	.19563014	.17147902	.15040221	.13199668	.11591372
29	.18455674	.16101316	.14056282	.12278761	.10732752
30	.17411013	.15118607	.13136712	.11422103	.09937733
31	.16425484	.14195875	.12277301	.10625212	.09201605
32	.15495740	.13329460	.11474113	.09883918	.08520005
33	.14618622	.12515925	.10723470	.09194343	.07888893
34	.13791153	.11752042	.10021934	.08552877	.07304531
35	.13010522	.11034781	.09366294	.07956164	.06763454
36	.12274077	.10361297	.08753546	.07401083	.06262458
37	.11579318	.09728917	.08180884	.06884729	.05798572
38	.10923885	.09135134	.07645686	.06404399	.05369048
39	.10305552	.08577590	.07145501	.05957580	.04971341
40	.09722219	.08054075	.06678038	.05541935	.04603093
41	.09171905	.07562512	.06241157	.05155288	.04262123
42	.08652740	.07100950	.05832857	.04795617	.03946411
43	.08162962	.06667559	.05451268	.04461039	.03654084
44	.07700908	.06260619	.05094643	.04149804	.03383411
45	.07265007	.05878515	.04761349	.03860283	.03132788
46	.06853781	.05519733	.04449859	.03590961	.02900730
47	.06465831	.05182848	.04158747	.03340428	.02685861
48	.06099840	.04866524	.03886679	.03107375	.02486908
49	.05754566	.04569506	.03632410	.02890582	.02302693
50	.05428836	.04290616	.03394776	.02688913	.02132123

CHAPTER 5

Annuities Certain

WHAT IS AN ANNUITY CERTAIN?

Suppose you buy a house with a $100,000 loan and promise to make monthly payments of interest and principal of $733.76 for 30 years. Your monthly payments form a stream of equal payments, at equal time intervals, for 360 months. Such a stream of payments is called an annuity certain.

Many readers may have parents who receive a pension, or you may receive a pension yourself. They (or you) could possibly receive a Social Security pension or a retirement pension from a former employer. These pensions are also called annuities, but they are not annuities certain because they have a life contingency feature. When the person receiving the pension dies, the pension stops or perhaps is paid to someone else at a reduced amount. These pensions are called life annuities. Later in the book, we'll study how to compute their values as well. However, an annuity certain has fixed payments at equally spaced time intervals, and the payments will be paid to somebody.

EXAMPLES OF ANNUITIES CERTAIN

You finally lease the car of your dreams. You have signed a 36-month lease, at $1,997.32 per month, on the Phantom Super Imperial Deluxe Year 2003 model. The 36-monthly payments of $1,997.32 form an annuity certain.

You have lent your neighbor $100. She promises to repay you at $10 per month for 10 months. The 10 monthly payments of $10 each form an annuity certain.

You buy a $1,000 bond issued by the XYZ Corporation. The bond has a 9% coupon rate and matures in 20 years. The coupon interest is paid semiannually. The 40 semiannual coupon payments of $45 form an annuity certain.

WHY ANNUITIES CERTAIN ARE IMPORTANT

Many such examples exist. Most people like to receive a regular income, and most people, both borrowers and lenders, like to have debts paid off in equal installments over the life of the original loan.

There are good business reasons for this preference. It is good loan management policy to pay off the loan in installments over the original life of the loan. For example, before the Great Depression of the 1930s, many homeowners borrowed money to buy their homes, just as they do now. In those days, however, usually the homeowner just paid the interest on the loan, frequently quarterly, and simply rolled over the mortgage loan when it matured. When the Depression hit, many of these people could no longer pay the interest and could not refinance their mortgages when they came due. As a result, they frequently lost their homes, which they possibly had been living in for many years. They would have been paying for years, and if they had paid down the mortgage gradually over the years, they would have kept their homes because the mortgage would have been paid off or would have had only a small balance remaining. This is one reason for the introduction of the monthly payment self-amortizing mortgages during the 1930s.

Many bond issuers have a sinking fund as part of the bond contract for their new bond issues. A sinking fund is a series of payments that will pay off all or part of a bond issue before final maturity, just as a homeowner's monthly payment pays off almost all of the mortgage before final maturity. Many business loans have a similar sinking fund feature.

Paying off loans in installments over the life of the loan is good financial policy, both for the borrower and for the lender. The annuity certain, as a means of repaying the loan, is frequently an easy and convenient way to do this.

Annuities certain are also easy to evaluate because the payments are the same size, they are made at regular time intervals, and the payments are all evaluated using the same interest rate. We'll get to evaluations using different interest rates later in the book.

Annuities certain occur very often in finance. Bond coupon payments, loan payments, annuity payments, and many other contractual payments all often form annuities certain. You should make sure you understand how annuities certain work and how to compute the mathematical present value of an annuity certain.

THE EQUATION FOR THE PRESENT VALUE OF AN ANNUITY CERTAIN

An annuity consists of a series of payments, starting at time $t = 1$ period and ending at time $t = n$ periods. This annuity will have n payments. We assume that each of the payments is 1. The annuity will therefore have n equal payments of 1, each one period apart from the previous and from the next payment, and starting in one period. For payments unequal to 1, simply multiply the present value of the annuity by the amount of the payment.

The present value of each payment, made at time j, will be v^j, with $j = 1$, ..., n, and v^j indicating v raised to the power j, equal to the present value of the payment of 1 due at time $t = j$.

Therefore, the present value of the annuity will be

$$a_{\overline{n}|} = v^1 + ... + v^n \quad \text{(summing the present value of the } n \text{ payments of 1)}$$

Equation 5.1

$$= v(1 + ... + v^{n-1})$$

Equation 5.2

Using the algebraic factorization

$$(a^n - b^n) = (a - b)(a^{n-1} + a^{n-2}b + ... + ab^{n-2} + b^{n-1})$$

If $a = 1$, and $b = a$, and dividing through by $(a - b)$ $(a \neq b$ and $1 \neq v)$, then

$$(1 + v + ... + v^{n-1}) = (1 - v^n)/(1 - v)$$

Equation 5.3

Substituting into Equation. 5.2, we have

$$a_{\overline{n}|} = \frac{v(1 - v^n)}{1 - v}$$

Equation 5.4

(Note: $a_{\overline{n}|}$ is called a angle n and indicates an annuity certain with n payments.)

Substituting $v = 1/(1 + i)$, we have

$$a_{\overline{n}|} = \frac{\left(\dfrac{1}{1+i}\right)\left(1 - \dfrac{1}{(1+i)^n}\right)}{\left(1 - \dfrac{1}{(1+i)}\right)} \qquad \text{Equation 5.5}$$

Multiplying numerator and denominator by $(1 + i)$ $(i \neq -1)$, we obtain

$$a_{\overline{n}|} = \frac{\left(1 - \dfrac{1}{(1+i)^n}\right)}{i} \qquad \text{Equation 5.6}$$

Equation 5.6 is the equation for the present value of an annuity of 1 for n periods, at interest rate i per period, $(i \neq -1)$, $(i \neq 0)$.

A LOOK AT THE TABLES FOR AN ANNUITY CERTAIN

Look at the table values for the present value of an annuity, as shown on pages 86–93. You can see that as the interest rate increases, the value of the annuity goes down. You need to put aside less money to buy the same payments because your money earns at a greater rate. As the number of periods increases, the value of the annuity increases. More money is always worth more money, even if the eventual payment is a long time off. However, the value increases at a decreasing rate, because the additional payments are due at increasingly distant times and have ever-decreasing present values.

SOLVING FOR THE INTEREST RATE, GIVEN THE ANNUITY CERTAIN AND ITS COST

As in most similar cases, you cannot usually solve the equation to find the interest rate if you have the annuity series and its value. Use the bisection method of solving, as discussed in Chapter 3.

THE PERPETUITY

An annuity that will continue without time limit (forever) is called a perpetuity. They rarely occur, although sometimes bonds have long maturity periods, such as the recent Walt Disney 100-year bonds.

The equation for a perpetuity of 1 can be expressed as follows:

$$1 = iP,$$

where   1 = the size of the annuity,
        i = the interest rate,
        P = the value of the perpetuity
Then $P = 1/i$, by simply dividing through by i

Not many perpetuities exist. The most famous were the old British Consols (consolidated loan), which guaranteed interest payments forever, although they were callable at par. The most recent issue was at $2\frac{1}{2}\%$, issued in 1888 to refund a previous issue of 3% Consols. Consols are mentioned in many 19th century works of English literature. The Canadian Pacific Railroad and the Canadian National Railroad both also had perpetual 4% outstanding at one time. I know of no American perpetuals, although during the 19th century, one American railroad issued bonds with a maturity year of 2862. In practical terms, this might as well have been a perpetual.

THE ANNUITY DUE

Occasionally you may come across the term "annuity due." This means an annuity with the first payment due right away, rather than one period in the future as is the case with the annuity certain. Clearly, an annuity due with n + 1 payments at a given interest rate is worth the value of an annuity (at the same rate) with n payments, plus 1, because of the payment of 1 made right away. The annuity due symbol is sometimes a with two dots over it ($\ddot{a}$) and called a double-dot. We won't use the annuity due concept in this book, but you should know what it is.

EXAMPLES

EXAMPLE 5.1.   Recall the college fund discussed in the previous chapter. You want to set up a college fund for your 2-year old child. In the previous chapter, we computed that you needed $39,364.63 now to have $100,000 when your child would be 18, assuming you can earn 6% in the meantime. You don't have $39,364.63 now, but you can make a series of annual payments. How much must you pay annually to have the $100,000 for your child?

The present value of the annuity is $39,364.63. The value of an annuity of 1, for 16 years at 6%, is 10.1058953 (taken from the interest tables). The value you need is $39,364.63. Dividing this by the value of an annuity of 1 (10.1058953) gives $3,895.21. This is the amount you must provide each year, starting in one year, to have $100,000 for your child in 16 years. This amount, about $325 monthly, is in the reasonable range for many people.   ■

EXAMPLE 5.2.   Recall the new car example from the previous chapter. You want to buy a new car. We computed that you need a present value of

$15,302.68 to have $20,000 in 5 years, assuming an interest rate of 5.5% true annual rate per year. You don't have this much now, but how much must you pay annually to have the $20,000 in 5 years to buy the car?

An annuity of 1 per year, for 5 years, at 5.5%, has a present value of 4.27028448 (taken from the tables). Your present value is $15,302.68. The annual payment you need to make is (15,302.68/4.27028448) = 3,583.53. You will need to make five annual payments of $3,583.53 (about $300 monthly) to buy your new car in five years. ■

EXAMPLE 5.3.   You have loaned your neighbor $100. Your neighbor will repay you at $10 per month, starting in 1 month. Assume an interest rate of 1% per month. The present value of an annuity of 1, for 10 payments at 1% per period, is 9.47130453. In this case, the present value of the actual annuity is ($10)(9.47130453) = $94.71. At 1% per month interest, the value of your loan to your neighbor is $94.71, and you gave her an extra $5.29 ($100 − $94.71) in value, at 1% per month interest rate.   ■

EXAMPLE 5.4.   The XYZ Corporation has a 9% bond, with a coupon payment of $45 every 6 months for 20 years. Here we have an annuity of $45 every 6 months for 20 years, or 40 payments. We need an interest rate to evaluate the annuity. Suppose we pick a rate of 4% every half year, for a nominal annual rate of 8%. From the table, the present value of an annuity of 1 at 4% for 40 periods is 19.7927739. The value of our coupon is 45 times this, or (45)(19.7927739) = 890.67.

In this case, the maturing principal also has a value. You will need to add this to figure the value of the bond. This process is discussed in the next chapter.   ■

FURTHER COMMENTS

If you are buying an annuity, the net price you will pay will be the present value of the annuity. Your outflow will be the cost of the annuity, and your inflows will be the payments you receive. The interest rate i is the rate that makes the present values of the outflows and the inflows equal. In the previous chapter, we called this the internal rate of return. In the case of an annuity certain, the rate you use to evaluate the present value of the annuity is the internal rate of return. In finance, this is also called the yield.

In this chapter, we computed the present values of the future payments using the same interest rate for all the payments. We don't need to use the same rate. We could use a different rate for some payments and could even use a different rate for each payment. Later in the book, we will demonstrate a way to obtain these different rates and use them to compute the present value of a future flow of funds.

But if you use more than one rate to evaluate the present value of the payments, or if you have payments of differing sizes, then you cannot use the equation we developed earlier in this chapter. That equation requires that all the payments be the same size, that the present values all be evaluated at the same interest rate, and that the time periods between payments all be the same. These conditions frequently happen in finance, which makes the present value of an annuity equation especially important.

ANALYSIS AND CALCULATION OF SOME COMBINATION ANNUITIES CERTAIN

Sometimes you may have a problem that requires you to calculate the present value of part of an annuity certain. You may be able to compute the required value by analyzing the flow of funds and splitting it up into several annuities certain or expressing it in some combination of annuities certain. Here are two examples.

EXAMPLE 5.5.   You are receiving an annuity of 2 for 10 years, and then an annuity of 1 for 10 additional years. Using $i = 6\%$, what is the value of your annuity?

Split your series of payments into an annuity of 1 for 10 years and an annuity of 1 for 20 years. Adding these two annuities together will give you your annuity. An annuity of 1 for 10 years at 6% is 7.36008705. An annuity of 1 for 20 years at 6% is 11.4699212 (both values from the table). Adding these together gives 18.83000825. This is the value of your annuity.   ■

EXAMPLE 5.6.   You will receive an annuity of 1 for 10 years starting 11 years from now and ending 20 years from now. Using a 6% interest rate, what is the present value of your annuity?

You can compute the present value by taking the value of a 20-year annuity and deducting from it the value of a 10-year annuity, representing the payments you will not be receiving. The 20-year value is 11.4699212, and the 10-year value is 7.36008705 (both values from the table). The value of your annuity is the 20-year annuity value minus the 10-year annuity value, or 11.4699212 − 7.36008705. This is 4.10983415, the value of your annuity.

There are no general rules for this kind of work. It depends on your own ingenuity.   ■

CHAPTER SUMMARY

An annuity certain is a series of equal payments, equally spaced in time, starting in one period.

Life annuities have a life contingency factor and are not annuities certain. The equation for an annuity certain with n payments is

$$a_{\overline{n}|} = \frac{\left[1 - \dfrac{1}{(1+i)^n}\right]}{i}, \qquad (i \neq -1), (i \neq 0)$$

To solve for the interest rate, given the annuity certain and its cost, use the bisection method discussed earlier or some other numerical analysis approach.

Annuities certain are important in finance, especially in figuring and scheduling loan repayments. Scheduled loan repayments are good financial policy for both borrower and lender.

For the annuity certain equation to hold, all the payments must be the same, they must all be evaluated at the same interest rate, and the time periods between the payments must all be the same.

COMPUTER PROJECT

For a range of interest rates, say 1% to 30%, compute the values of a perpetuity, an annuity certain for 100 years, and the difference. How much would you pay to convert a 100-year annuity certain to a perpetuity? What is the difference between the calculated value and the amount you would pay? What, if anything, does this tell you about the psychology of markets?

TOPICS FOR CLASS DISCUSSION

1. You are retiring at age 60 with a pension of $1,500 monthly. You can receive a Social Security pension of $1,100 monthly, starting in 2 years. Your human resources department has offered to pay you an increased pension for 2 years, until your Social Security payments start, and then a reduced pension. They will pay you $2,300 for 2 years, and then a lifetime pension of $1,200. This $1,200, combined with Social Security's $1,100, will give you $2,300 monthly. What factors should you consider in deciding whether or not to accept this offer?

2. You are now 62 and can retire with a Social Security pension. If you wait until age 65, you will receive Social Security of $1,100 monthly. If you retire now, you will receive Social Security of $880 monthly (80%

of the age 65 pension). What factors should you consider in making this decision?

3. You are borrowing $100,000 to buy a home. The bank will lend you the full $100,000 for 30 years at a 6% nominal annual rate. It will also lend you $99,000 for 30 years at a 5.75% nominal annual rate. (This is called paying points. In this case, you are paying one point [1%, or $1,000], but you will make monthly payments, and other repayments, based on the full $100,000.) What factors should you consider in deciding which offer to accept?

PROBLEMS

1. You are borrowing on a mortgage to buy a home. The bank will offer you $100,000 at 6% with no points, at 5.75% with one point, and at 5.5% with two points. You intend to stay in the house for many years. You can afford any of the three plans and are only interested in reducing your interest cost. Which plan should you select?
2. You are buying a U.S. Treasury bond with a 6% coupon, due in 25 years. Treasuries of that maturity are yielding 5% interest. What is the value of the coupon stream you will receive?
3. Yields on Treasuries have just fallen sharply, and the bond in Problem 2 above now yields 4.75%. What is the new value of the coupons you will receive?
4. You need to put aside enough to have $120,000 to pay for a college education for your newborn child. You think you can earn 7% annually in various investments over that period. How much must you put aside annually to have that amount when your child goes off to college at age 18?
5. You will receive payments each year of $5,000, starting at 11 years and continuing for 20 years. Assume an interest rate of 6.5%. How much is this future income stream worth now?
6. What is the equation for an annuity certain?
7. State how the equation for an annuity certain is developed.

INTEREST TABLES (Continued)

PRESENT VALUE OF ANNUITY $[1 - (1 + i)^{-n}]/i$

The following table gives the present value of an annuity of unit value per period for a term of n periods at rate of interest i per period; usually indicated as $a_{\overline{n}|}$ at i.

Years n	Rate i				
	.0025($\frac{1}{4}$%)	.004167($\frac{5}{12}$%)	.005($\frac{1}{2}$%)	.005833($\frac{7}{12}$%)	.0075($\frac{3}{4}$%)
1	0.99750623	0.99585062	0.99502488	0.99420050	0.99255583
2	1.99252492	1.98756908	1.98509938	1.98263513	1.97772291
3	2.98506227	2.97517253	2.97024814	2.96533732	2.95555624
4	3.97512446	3.95867804	3.95049566	3.94234034	3.92611041
5	4.96271766	4.93810261	4.92586633	4.91367722	4.88943961
6	5.94784804	5.91346318	5.89638441	5.87938083	5.84559763
7	6.93052174	6.88477661	6.86207404	6.83948384	6.79463785
8	7.91074487	7.85205970	7.82295924	7.79401874	7.73661325
9	8.88852357	8.81532916	8.77906392	8.74301780	8.67157642
10	9.86386391	9.77460165	9.73041186	9.68651314	9.59957958
11	10.8367720	10.7298938	10.6770267	10.6245367	10.5206745
12	11.8072538	11.6812220	11.6189321	11.5571201	11.4349127
13	12.7753156	12.6286028	12.5561513	12.4842951	12.3423451
14	13.7409631	13.5720526	13.4887078	13.4060929	13.2430224
15	14.7042026	14.5115877	14.4166246	14.3225447	14.1369950
16	15.6650400	15.4472242	15.3399250	15.2336816	15.0243126
17	16.6234813	16.3789785	16.2586319	16.1395343	15.9050249
18	17.5795325	17.3068665	17.1727680	17.0401335	16.7791811
19	18.5331995	18.2309044	18.0823562	17.9355097	17.6468298
20	19.4844883	19.1511082	18.9874191	18.8256931	18.5080197
21	20.4334048	20.0674936	19.8879793	19.7107140	19.3627987
22	21.3799549	20.9800766	20.7840590	20.5906021	20.2112146
23	22.3241445	21.8888730	21.6756806	21.4653874	21.0533147
24	23.2659796	22.7938984	22.5628662	22.3350993	21.8891461
25	24.2054659	23.6951685	23.4456380	23.1997673	22.7187555
26	25.1426094	24.5926989	24.3240179	24.0594207	23.5421891
27	26.0774158	25.4865052	25.1980278	24.9140885	24.3594929
28	27.0098911	26.3766027	26.0676894	25.7637997	25.1707125
29	27.9400410	27.2630068	26.9330242	26.6085830	25.9758933
30	28.8678713	28.1457329	27.7940540	27.4484669	26.7750802
31	29.7933879	29.0247963	28.6508000	28.2834799	27.5683178
32	30.7165964	29.9002120	29.5032835	29.1136503	28.3556504
33	31.6375026	30.7719954	30.3515259	29.9390061	29.1371220
34	32.5561123	31.6401614	31.1955482	30.7595752	29.9127762
35	33.4724313	32.5047250	32.0353713	31.5753855	30.6826563
36	34.3864651	33.3657013	32.8710162	32.3864645	31.4468053
37	35.2982196	34.2231050	33.7025037	33.1928396	32.2052658
38	36.2077003	35.0769511	34.5298544	33.9945381	32.9580802
39	37.1149130	35.9272542	35.3530890	34.7915872	33.7052905
40	38.0198634	36.7740290	36.1722279	35.5840137	34.4469384
41	38.9225570	37.6172903	36.9872914	36.3718446	35.1830654
42	39.8229995	38.4570526	37.7982999	37.1551065	35.9137126
43	40.7211965	39.2933304	38.6052735	37.9338259	36.6389207
44	41.6171536	40.1261382	39.4082324	38.7080290	37.3587302
45	42.5108764	40.9554903	40.2071964	39.4777422	38.0731814
46	43.4023705	41.7814011	41.0021855	40.2429914	38.7823140
47	44.2916414	42.6038849	41.7932194	41.0038026	39.4861677
48	45.1786946	43.4229559	42.5803178	41.7602014	40.1847819
49	46.0635358	44.2386283	43.3635003	42.5122135	40.8781954
50	46.9461704	45.0509162	44.1427864	43.2598643	41.5664471

PRESENT VALUE OF ANNUITY $[1 - (1 + i)^{-n}]/i$
(Continued)

Years	Rate i				
n	.0025($\frac{1}{4}$%)	.004167($\frac{5}{12}$%)	.005($\frac{1}{2}$%)	.005833($\frac{7}{12}$%)	.0075($\frac{3}{4}$%)
50	46.9461704	45.0509162	44.1427864	43.2598643	41.5664471
51	47.8266039	45.8598335	44.9181954	44.0031791	42.2495753
52	48.7048418	46.6653944	45.6897466	44.7421830	42.9276181
53	49.5808895	47.4676127	46.4574593	45.4769011	43.6006135
54	50.4547527	48.2665022	47.2213526	46.2073582	44.2685990
55	51.3264366	49.0620769	47.9814454	46.9335789	44.9316119
56	52.1959467	49.8543505	48.7377566	47.6555880	45.5896893
57	53.0632885	50.6433366	49.4903050	48.3734098	46.2428678
58	53.9284673	51.4290489	50.2391095	49.0870686	46.8911839
59	54.7914886	52.2115009	50.9841886	49.7965885	47.5346738
60	55.6523577	52.9907063	51.7255608	50.5019935	48.1733735
61	56.5110800	53.7666785	52.4632445	51.2033075	48.8073186
62	57.3676608	54.5394309	53.1972582	51.9005543	49.4365445
63	58.2221056	55.3089768	53.9276201	52.5937574	50.0610864
64	59.0744195	56.0753296	54.6543484	53.2829402	50.6809791
65	59.9246080	56.8385025	55.3774611	53.9681262	51.2962571
66	60.7726763	57.5985087	56.0969762	54.6493384	51.9069550
67	61.6186297	58.3553614	56.8129117	55.3265999	52.5131067
68	62.4624736	59.1090736	57.5252852	55.9999336	53.1147461
69	63.3042130	59.8596583	58.2341147	56.6693623	53.7119068
70	64.1438534	60.6071286	58.9394176	57.3349087	54.3046221
71	64.9813999	61.3514974	59.6412115	57.9965952	54.8929252
72	65.8168597	62.0927775	60.3395139	58.6544443	55.4768488
73	66.6502322	62.8309817	61.0343422	59.3084781	56.0564256
74	67.4815283	63.5661229	61.7257137	59.9587190	56.6316879
75	68.3107515	64.2982136	62.4136454	60.6051887	57.2026679
76	69.1379067	65.0272667	63.0981547	61.2479092	57.7693975
77	69.9629992	65.7532946	63.7792584	61.8869023	58.3319081
78	70.7860341	66.4763100	64.4569735	62.5221895	58.8902314
79	71.6070166	67.1963253	65.1313169	63.1537924	59.4443984
80	72.4259517	67.9133530	65.8023054	63.7817323	59.9944401
81	73.2428446	68.6274055	66.4699556	64.4060304	60.5403872
82	74.0577003	69.3384951	67.1342842	65.0267080	61.0822702
83	74.8705240	70.0466341	67.7953076	65.6437859	61.6201193
84	75.6813207	70.7518348	68.4530424	66.2572851	62.1539646
85	76.4900955	71.4541094	69.1075049	66.8672262	62.6838358
86	77.2968533	72.1534699	69.7587114	67.4736301	63.2097626
87	78.1015993	72.8499285	70.4066780	68.0765171	63.7317743
88	78.9043385	73.5434973	71.0514209	68.6759076	64.2499000
89	79.7050758	74.2341882	71.6929561	69.2718220	64.7641688
90	80.5038163	74.9220131	72.3312996	69.8642803	65.2746092
91	81.3005649	75.6069840	72.9664672	70.4533027	65.7812498
92	82.0953265	76.2891127	73.5984749	71.0389091	66.2841189
93	82.8881003	76.9684110	74.2273382	71.6211192	66.7832446
94	83.6789090	77.6448906	74.8530728	72.1999528	67.2786547
95	84.4677397	78.3185633	75.4756943	72.7754295	67.7703768
96	85.2546031	78.9894406	76.0952183	73.3475687	68.2584386
97	86.0395044	79.6575342	76.7116600	73.9163897	68.7428671
98	86.8224483	80.3228557	77.3250348	74.4819119	69.2236894
99	87.6034397	80.9854164	77.9353580	75.0441544	69.7009324
100	88.3824835	81.6452280	78.5426448	75.6031361	70.1746227

INTEREST TABLES (Continued)

PRESENT VALUE OF ANNUITY $[1 - (1 + i)^{-n}]/i$
(Continued)

Years n	Rate i				
	.01(1 %)	.01125(1⅛ %)	.0125(1¼ %)	.015(1½ %)	.0175(1¾ %)
1	0.99009901	0.98887515	0.98765432	0.98522167	0.98280098
2	1.97039506	1.96674923	1.96311538	1.95588342	1.94869875
3	2.94098521	2.93374460	2.92653371	2.91220042	2.89798403
4	3.90196555	3.88998230	3.87805798	3.85438465	3.83094254
5	4.85343124	4.83558200	4.81783504	4.78264497	4.74785508
6	5.79547647	5.77066205	5.74600992	5.69718717	5.64899762
7	6.72819453	6.69533948	6.66272585	6.59821396	6.53464139
8	7.65167775	7.60973002	7.56812429	7.48592508	7.40505297
9	8.56601758	8.51394810	8.46234498	8.36051732	8.26049432
10	9.47130453	9.40810690	9.34552591	9.22218455	9.10122291
11	10.3676282	10.2923183	10.2178034	10.0711178	9.92749181
12	11.2550775	11.1666930	11.0793120	10.9075052	10.7395497
13	12.1337401	12.0313404	11.9301847	11.7315322	11.5376410
14	13.0037030	12.8863688	12.7705527	12.5433815	12.3220059
15	13.8650525	13.7318851	13.6005459	13.3432330	13.0928805
16	14.7178738	14.5679951	14.4202923	14.1312640	13.8504968
17	15.5622513	15.3948036	15.2299183	14.9076493	14.5950828
18	16.3982686	16.2124139	16.0295489	15.6725609	15.3268627
19	17.2260085	17.0209285	16.8193076	16.4261684	16.0460567
20	18.0455530	17.8204485	17.5993161	17.1686388	16.7528813
21	18.8569831	18.6110739	18.3696949	17.9001367	17.4475492
22	19.6603703	19.3929037	19.1305629	18.6208244	18.1302695
23	20.4558211	20.1660358	19.8820374	19.3308614	18.8012476
24	21.2433873	20.9305669	20.6242345	20.0304054	19.4606856
25	22.0231557	21.6865928	21.3572687	20.7196112	20.1087820
26	22.7952037	22.4342079	22.0812530	21.3986317	20.7457317
27	23.5596076	23.1735060	22.7962993	22.0676175	21.3717264
28	24.3164432	23.9045795	23.5025178	22.7267167	21.9869547
29	25.0657853	24.6275199	24.2000176	23.3760756	22.5916017
30	25.8077082	25.3424177	24.8889062	24.0158380	23.1858493
31	26.5422854	26.0493623	25.5692901	24.6461458	23.7698765
32	27.2695895	26.7484424	26.2412742	25.2671387	24.3438590
33	27.9896925	27.4397452	26.9049622	25.8789544	24.9079695
34	28.7026659	28.1233575	27.5604564	26.4817285	25.4623779
35	29.4085801	28.7993646	28.2078582	27.0755946	26.0072510
36	30.1075050	29.4678513	28.8472674	27.6606843	26.5427528
37	30.7995099	30.1289011	29.4787826	28.2371274	27.0690445
38	31.4846633	30.7825969	30.1025013	28.8050516	27.5862846
39	32.1630330	31.4290204	30.7185198	29.3645829	28.0946286
40	32.8346861	32.0682526	31.3269332	29.9158452	28.5942295
41	33.4996892	32.7003734	31.9278352	30.4589608	29.0852379
42	34.1581081	33.3254620	32.5213187	30.9940500	29.5678014
43	34.8100081	33.9435965	33.1074753	31.5212316	30.0420652
44	35.4554535	34.5548544	33.6863954	32.0406222	30.5081722
45	36.0945084	35.1593121	34.2581683	32.5523372	30.9662626
46	36.7272361	35.7570454	34.8228822	33.0564898	31.4164743
47	37.3536991	36.3481289	35.3806244	33.5531920	31.8589428
48	37.9739595	36.9326367	35.9314809	34.0425536	32.2938013
49	38.5880787	37.5106420	36.4755367	34.5246834	32.7211806
50	39.1961175	38.0822171	37.0128758	34.9996881	33.1412095

INTEREST TABLES (Continued)

PRESENT VALUE OF ANNUITY $[1 - (1 + i)^{-n}]/i$
(Continued)

Years n	.01(1%)	.01125(1⅛%)	.0125(1¼%)	.015(1½%)	.0175(1¾%)
			Rate i		
50	39.1961175	38.0822171	37.0128758	34.9996881	33.1412095
51	39.7981362	38.6474335	37.5435810	35.4676730	33.5540142
52	40.3941942	39.2063619	38.0677343	35.9287419	33.9597191
53	40.9843507	39.7590723	38.5854166	36.3829969	34.3584463
54	41.5686641	40.3056339	39.0967078	36.8305388	34.7503158
55	42.1471922	40.8461151	39.6016867	37.2714668	35.1354455
56	42.7199922	41.3805836	40.1004313	37.7058786	35.5139513
57	43.2871210	41.9091061	40.5930186	38.1338706	35.8859473
58	43.8486347	42.4317490	41.0795245	38.5555375	36.2515452
59	44.4045888	42.9485775	41.5600242	38.9709729	36.6108553
60	44.9550384	43.4596563	42.0345918	39.3802689	36.9639855
61	45.5000380	43.9650495	42.5033005	39.7835161	37.3110423
62	46.0396416	44.4648203	42.9662228	40.1808041	37.6521300
63	46.5739026	44.9590312	43.4234299	40.5722208	37.9873514
64	47.1028738	45.4477441	43.8749925	40.9578530	38.3168072
65	47.6266078	45.9310201	44.3209802	41.3377862	38.6405968
66	48.1451562	46.4089197	44.7614619	41.7121046	38.9588175
67	48.6585705	46.8815028	45.1965056	42.0808912	39.2715651
68	49.1669015	47.3488285	45.6261784	42.4442278	39.5789337
69	49.6701995	47.8109553	46.0505466	42.8021949	39.8810160
70	50.1685143	48.2679409	46.4696756	43.1548718	40.1779027
71	50.6618954	48.7198427	46.8836302	43.5023368	40.4696832
72	51.1503915	49.1667171	47.2924743	43.8446668	40.7564454
73	51.6340510	49.6086202	47.6962709	44.1819377	41.0382766
74	52.1129218	50.0456071	48.0950824	44.5142243	41.3152586
75	52.5870512	50.4777326	48.4889703	44.8416003	41.5874777
76	53.0564864	50.9050508	48.8779953	45.1641383	41.8550149
77	53.5212736	51.3276151	49.2622176	45.4819096	42.1179508
78	53.9814590	51.7454785	49.6416964	45.7949848	42.3763644
79	54.4370882	52.1586932	50.0164903	46.1034333	42.6303336
80	54.8882061	52.5673109	50.3866571	46.4073235	42.8799347
81	55.3348575	52.9713829	50.7522539	46.7067227	43.1252430
82	55.7770867	53.3709596	51.1133372	47.0016972	43.3663322
83	56.2149373	53.7660910	51.4699626	47.2923125	43.6032749
84	56.6484528	54.1568267	51.8221853	47.5786330	43.8361424
85	57.0776760	54.5432156	52.1700596	47.8607222	44.0650048
86	57.5026495	54.9253059	52.5136391	48.1386425	44.2899310
87	57.9234154	55.3031455	52.8529769	48.4124557	44.5109887
88	58.3400152	55.6767817	53.1881253	48.6822224	44.7282444
89	58.7524903	56.0462613	53.5191361	48.9480023	44.9417636
90	59.1608815	56.4116304	53.8460604	49.2098545	45.1516104
91	59.5652292	56.7729349	54.1689485	49.4678370	45.3578480
92	59.9655735	57.1302199	54.4878504	49.7220069	45.5605386
93	60.3619539	57.4835302	54.8028152	49.9724206	45.7597431
94	60.7544098	57.8329100	55.1138915	50.2191335	45.9555215
95	61.1429800	58.1784029	55.4211274	50.4622005	46.1479327
96	61.5277030	58.5200523	55.7245703	50.7016754	46.3370345
97	61.9086168	58.8579010	56.0242670	50.9376112	46.5228841
98	62.2857592	59.1919911	56.3202637	51.1700603	46.7055372
99	62.6591676	59.5223645	56.6126061	51.3990742	46.8850488
100	63.0288788	59.8490625	56.9013394	51.6247037	47.0614730

PRESENT VALUE OF ANNUITY $[1 - (1 + i)^{-n}]/i$
(Continued)

Years n	Rate i				
	.02(2 %)	.0225(2¼ %)	.025(2½ %)	.0275(2¾ %)	.03(3 %)
1	0.98039216	0.97799511	0.97560976	0.97323601	0.97087379
2	1.94156094	1.93446955	1.92742415	1.92042434	1.91346970
3	2.88388327	2.86989687	2.85602356	2.84226213	2.82861135
4	3.80772870	3.78474021	3.76197421	3.73942787	3.71709840
5	4.71345951	4.67945253	4.64582850	4.61258186	4.57970719
6	5.60143089	5.55447680	5.50812536	5.46236678	5.41719144
7	6.47199107	6.41024626	6.34939060	6.28940806	6.23028296
8	7.32548144	7.24718461	7.17013717	7.09431441	7.01969219
9	8.16223671	8.06570622	7.97086553	7.87767826	7.78610892
10	8.98258501	8.86621635	8.75206393	8.64007616	8.53020284
11	9.78684805	9.64911134	9.51420871	9.38206926	9.25262411
12	10.5753412	10.4147788	10.2577646	10.1042037	9.95400399
13	11.3483737	11.1635979	10.9831850	10.8070109	10.6349553
14	12.1062488	11.8959392	11.6909122	11.4910081	11.2960731
15	12.8492635	12.6121655	12.3813777	12.1566989	11.9379351
16	13.5777093	13.3126313	13.0550027	12.8045732	12.5611020
17	14.2918719	13.9976834	13.7121977	13.4351077	13.1661185
18	14.9920313	14.6676611	14.3533636	14.0487666	13.7535131
19	15.6784620	15.3228959	14.9788913	14.6460016	14.3237991
20	16.3514333	15.9637124	15.5891623	15.2272521	14.8774749
21	17.0112092	16.5904277	16.1845486	15.7929461	15.4150241
22	17.6580482	17.2033523	16.7654132	16.3434999	15.9369166
23	18.2922041	17.8027896	17.3321105	16.8793186	16.4436084
24	18.9139256	18.3890362	17.8849858	17.4007967	16.9355421
25	19.5234565	18.9623826	18.4243764	17.9083180	17.4131477
26	20.1210358	19.5231126	18.9506111	18.4022559	17.8768424
27	20.7068978	20.0715038	19.4640109	18.8829741	18.3270315
28	21.2812724	20.6078276	19.9648887	19.3508264	18.7641082
29	21.8443847	21.1323498	20.4535499	19.8061571	19.1884546
30	22.3964556	21.6453298	20.9302926	20.2493013	19.6004413
31	22.9377015	22.1470219	21.3954074	20.6805852	20.0004285
32	23.4683348	22.6376742	21.8491780	21.1003262	20.3887655
33	23.9885636	23.1175298	22.2918809	21.5088333	20.7657918
34	24.4985917	23.5868262	22.7237863	21.9064071	21.1318367
35	24.9986193	24.0457958	23.1451573	22.2933403	21.4872201
36	25.4888425	24.4946658	23.5562511	22.6699175	21.8322525
37	25.9694534	24.9336585	23.9573181	23.0364161	22.1672354
38	26.4406406	25.3629912	24.3486030	23.3931057	22.4924616
39	26.9025888	25.7828765	24.7303444	23.7402488	22.8082151
40	27.3554792	26.1935222	25.1027751	24.0781011	23.1147720
41	27.7994895	26.5951317	25.4661220	24.4069110	23.4124000
42	28.2347936	26.9879039	25.8206068	24.7269207	23.7013592
43	28.6615623	27.3720332	26.1664457	25.0383656	23.9819021
44	29.0799631	27.7477097	26.5038495	25.3414751	24.2542739
45	29.4901599	28.1151195	26.8330239	25.6364721	24.5187125
46	29.8923136	28.4744445	27.1541696	25.9235738	24.7754491
47	30.2865820	28.8258626	27.4674826	26.2029915	25.0247078
48	30.6731196	29.1695478	27.7731537	26.4749309	25.2667066
49	31.0520780	29.5056702	28.0713695	26.7395922	25.5016569
50	31.4236059	29.8343963	28.3623117	26.9971700	25.7297640

PRESENT VALUE OF ANNUITY $[1 - (1 + i)^{-n}]/i$
(Continued)

Years n	Rate i				
	.02(2 %)	.0225(2¼ %)	.025(2½ %)	.0275(2¾ %)	.03(3 %)
50	31.4236059	29.8343963	28.3623117	26.9971700	25.7297640
51	31.7878489	30.1558888	28.6461577	27.2478540	25.9512272
52	32.1449499	30.4703069	28.9230807	27.4918287	26.1662400
53	32.4950489	30.7778062	29.1932495	27.7292737	26.3749903
54	32.8382833	31.0785391	29.4568288	27.9603637	26.5776605
55	33.1747875	31.3726544	29.7139793	28.1852688	26.7744276
56	33.5046936	31.6602977	29.9648578	28.4041545	26.9654637
57	33.8281310	31.9416114	30.2096174	28.6171820	27.1509357
58	34.1452265	32.2167349	30.4484072	28.8245081	27.3310055
59	34.4561044	32.4858043	30.6813729	29.0262852	27.5058306
60	34.7608867	32.7489529	30.9086565	29.2226620	27.6755637
61	35.0596928	33.0063109	31.1303966	29.4137830	27.8403531
62	35.3526400	33.2580057	31.3467284	29.5997888	28.0003428
63	35.6398432	33.5041621	31.5577838	29.7808163	28.1556726
64	35.9214149	33.7449018	31.7636915	29.9569989	28.3064783
65	36.1974655	33.9803440	31.9645771	30.1284661	28.4528915
66	36.4681035	34.2106054	32.1605630	30.2953441	28.5950403
67	36.7334348	34.4357999	32.3517688	30.4577558	28.7330488
68	36.9935635	34.6560391	32.5383110	30.6158207	28.8670377
69	37.2485917	34.8714318	32.7203034	30.7696552	28.9971240
70	37.4986193	35.0820849	32.8978570	30.9193725	29.1234214
71	37.7437444	35.2881026	33.0710800	31.0650827	29.2460401
72	37.9840631	35.4895869	33.2400780	31.2068931	29.3650875
73	38.2196697	35.6866376	33.4049542	31.3449082	29.4806675
74	38.4506566	35.8793521	33.5658089	31.4792294	29.5928811
75	38.6771143	36.0678261	33.7227404	31.6099556	29.7018263
76	38.8991317	36.2521526	33.8758443	31.7371830	29.8075983
77	39.1167968	36.4324231	34.0252140	31.8610054	29.9102896
78	39.3301919	36.6087267	34.1709405	31.9815138	30.0099899
79	39.5394039	36.7811509	34.3131127	32.0987969	30.1067863
80	39.7445136	36.9497808	34.4518172	32.2129410	30.2007634
81	39.9456016	37.1147000	34.5871388	32.3240301	30.2920033
82	40.1427466	37.2759903	34.7191598	32.4321461	30.3805858
83	40.3360261	37.4337313	34.8479607	32.5373685	30.4665881
84	40.5255158	37.5880013	34.9736202	32.6397747	30.5500856
85	40.7112900	37.7388765	35.0962149	32.7394401	30.6311510
86	40.8934216	37.8864318	35.2158194	32.8364380	30.7098554
87	41.0719819	38.0307402	35.3325067	32.9308399	30.7862673
88	41.2470411	38.1718730	35.4463480	33.0227153	30.8604537
89	41.4186677	38.3099003	35.5574127	33.1121317	30.9324794
90	41.5869292	38.4448902	35.6657685	33.1991549	31.0024071
91	41.7518913	38.5769098	35.7714814	33.2838490	31.0702982
92	41.9136190	38.7060242	35.8746160	33.3662764	31.1362118
93	42.0721754	38.8322975	35.9752352	33.4464978	31.2002057
94	42.2276230	38.9557922	36.0734002	33.5245720	31.2623356
95	42.3800225	39.0765694	36.1691709	33.6005567	31.3226559
96	42.5294339	39.1946889	36.2626057	33.6745078	31.3812193
97	42.6759155	39.3102092	36.3537617	33.7464796	31.4380770
98	42.8195250	39.4231875	36.4426943	33.8165251	31.4932787
99	42.9603187	39.5336797	36.5294579	33.8846960	31.5468725
100	43.0983516	39.6417405	36.6141053	33.9510423	31.5989053

PRESENT VALUE OF ANNUITY $[1 - (1 + i)^{-n}]/i$
(Continued)

Years n	.035(3½%)	.04(4%)	.045(4½%)	.05(5%)	.055(5½%)
1	0.96618357	0.96153846	0.95693780	0.95238095	0.94786730
2	1.89969428	1.88609467	1.87266775	1.85941043	1.84631971
3	2.80163698	2.77509103	2.74896435	2.72324803	2.69793338
4	3.67307921	3.62989522	3.58752570	3.54595050	3.50515012
5	4.51505238	4.45182233	4.38997674	4.32947667	4.27028448
6	5.32855302	5.24213686	5.15787248	5.07569207	4.99553031
7	6.11454398	6.00205467	5.89270094	5.78637340	5.68296712
8	6.87395554	6.73274487	6.59588607	6.46321276	6.33456599
9	7.60768651	7.43533161	7.26879050	7.10782168	6.95219525
10	8.31660532	8.11089578	7.91271818	7.72173493	7.53762583
11	9.00155104	8.76047671	8.52891692	8.30641422	8.09253633
12	9.66333433	9.38507376	9.11858078	8.86325164	8.61851785
13	10.3027385	9.98564785	9.68285242	9.39357299	9.11707853
14	10.9205203	10.5631229	10.2228253	9.89864094	9.58964790
15	11.5174109	11.1183874	10.7395457	10.3796580	10.0375809
16	12.0941168	11.6522956	11.2340150	10.8377696	10.4621620
17	12.6513206	12.1656689	11.7071914	11.2740662	10.8646086
18	13.1896817	12.6592970	12.1599918	11.6895869	11.2460745
19	13.7098374	13.1339394	12.5932936	12.0853209	11.6076535
20	14.2124033	13.5903263	13.0079365	12.4622103	11.9503825
21	14.6979742	14.0291599	13.4047239	12.8211527	12.2752441
22	15.1671248	14.4511153	13.7844248	13.1630026	12.5831697
23	15.6204105	14.8568417	14.1477749	13.4885739	12.8750424
24	16.0583676	15.2469631	14.4954784	13.7986418	13.1516990
25	16.4815146	15.6220799	14.8282090	14.0939446	13.4139327
26	16.8903523	15.9827692	15.1466114	14.3751853	13.6624954
27	17.2853645	16.3295857	15.4513028	14.6430336	13.8980999
28	17.6670188	16.6630632	15.7428735	14.8981273	14.1214217
29	18.0357670	16.9837146	16.0218885	15.1410736	14.3331012
30	18.3920454	17.2920333	16.2888885	15.3724510	14.5337452
31	18.7362758	17.5884936	16.5443910	15.5928105	14.7239291
32	19.0688655	17.8735515	16.7888909	15.8026767	14.9041982
33	19.3902082	18.1476457	17.0228621	16.0025492	15.0750694
34	19.7006842	18.4111978	17.2467580	16.1929040	15.2370326
35	20.0006611	18.6646132	17.4610124	16.3741943	15.3905522
36	20.2904938	18.9082820	17.6660406	16.5468517	15.5360684
37	20.5705254	19.1425788	17.8622398	16.7112873	15.6739985
38	20.8410874	19.3678642	18.0499902	16.8678927	15.8047379
39	21.1024999	19.5844848	18.2296557	17.0170407	15.9286615
40	21.3550723	19.7927739	18.4015844	17.1590864	16.0461247
41	21.5991037	19.9930518	18.5661095	17.2943680	16.1574642
42	21.8348828	20.1856267	18.7235498	17.4232076	16.2629992
43	22.0626887	20.3707949	18.8742103	17.5459120	16.3630324
44	22.2827910	20.5488413	19.0183831	17.6627733	16.4578506
45	22.4954503	20.7200397	19.1563474	17.7740698	16.5477257
46	22.7009181	20.8846536	19.2883707	17.8800665	16.6329154
47	22.8994378	21.0429361	19.4147088	17.9810157	16.7136639
48	23.0912443	21.1951309	19.5356065	18.0771578	16.7902027
49	23.2765645	21.3414720	19.6512981	18.1687217	16.8627514
50	23.4556179	21.4821846	19.7620078	18.2559255	16.9315179

INTEREST TABLES (Continued)

PRESENT VALUE OF ANNUITY $[1 - (1 + i)^{-n}]/i$
(Continued)

Years n	Rate i .06(6 %)	.065(6½ %)	.07(7 %)	.075(7½ %)	.08(8 %)
1	0.94339623	0.93896714	0.93457944	0.93023256	0.92592593
2	1.83339267	1.82062642	1.80801817	1.79556517	1.78326475
3	2.67301195	2.64847551	2.62431604	2.60052574	2.57709699
4	3.46510561	3.42579860	3.38721126	3.34932627	3.31212684
5	4.21236379	4.15567944	4.10019744	4.04588490	3.99271004
6	4.91732433	4.84101356	4.76653966	4.69384642	4.62287966
7	5.58238144	5.48451977	5.38928940	5.29660132	5.20637006
8	6.20979381	6.08875096	5.97129851	5.85730355	5.74663894
9	6.80169227	6.65610419	6.51523225	6.37888703	6.24688791
10	7.36008705	7.18883022	7.02358154	6.86408096	6.71008140
11	7.88687458	7.68904246	7.49867434	7.31542415	7.13896426
12	8.38384394	8.15872532	7.94268630	7.73527827	7.53607802
13	8.85268296	8.59974208	8.35765074	8.12584026	7.90377594
14	9.29498393	9.01384233	8.74546799	8.48915373	8.24423698
15	9.71224899	9.40266885	9.10791401	8.82711975	8.55947869
16	10.1058953	9.76776418	9.44664860	9.14150674	8.85136916
17	10.4772597	10.1105767	9.76322299	9.43395976	9.12163811
18	10.8276035	10.4324664	10.0590869	9.70600908	9.37188714
19	11.1581165	10.7347102	10.3355952	9.95907821	9.60359920
20	11.4699212	11.0185072	10.5940142	10.1944914	9.81814741
21	11.7640766	11.2849833	10.8355273	10.4134803	10.0168032
22	12.0415817	11.5351956	11.0612405	10.6171910	10.2007437
23	12.3033790	11.7701367	11.2721874	10.8066893	10.3710589
24	12.5503575	11.9907387	11.4693340	10.9829668	10.5287583
25	12.7833562	12.1978767	11.6535832	11.1469459	10.6747762
26	13.0031662	12.3923725	11.8257787	11.2994845	10.8099780
27	13.2105341	12.5749977	11.9867090	11.4413810	10.9351648
28	13.4061643	12.7464767	12.1371113	11.5733776	11.0510785
29	13.5907210	12.9074898	12.2776741	11.6961652	11.1584060
30	13.7648312	13.0586759	12.4090412	11.8103863	11.2577833
31	13.9290860	13.2006347	12.5318142	11.9166384	11.3497994
32	14.0840434	13.3339293	12.6465553	12.0154776	11.4349994
33	14.2302296	13.4590885	12.7537900	12.1074210	11.5138884
34	14.3681411	13.5766089	12.8540094	12.1929498	11.5869337
35	14.4982464	13.6869567	12.9476723	12.2725114	11.6545682
36	14.6209871	13.7905697	13.0352078	12.3465222	11.7171928
37	14.7367803	13.8878589	13.1170166	12.4153695	11.7751785
38	14.8460192	13.9792102	13.1934735	12.4794135	11.8288690
39	14.9490747	14.0649861	13.2649285	12.5389893	11.8785824
40	15.0462969	14.1455269	13.3317088	12.5944087	11.9246133
41	15.1380159	14.2211520	13.3941204	12.6459615	11.9672346
42	15.2245433	14.2921615	13.4524490	12.6939177	12.0066987
43	15.3061729	14.3588371	13.5069617	12.7385281	12.0432395
44	15.3831820	14.4214433	13.5579081	12.7800261	12.0770736
45	15.4558321	'14.4802284	13.6055216	12.8186290	12.1084015
46	15.5243699	14.5354257	13.6500202	12.8545386	12.1374088
47	15.5890282	14.5872542	13.6916076	12.8879429	12.1642674
48	15.6500266	14.6359195	13.7304744	12.9190166	12.1891365
49	15.7075723	14.6816145	13.7667985	12.9479224	12.2121634
50	15.7618606	14.7245207	13.8007463	12.9748116	12.2334846

ANNUITY WHOSE PRESENT VALUE IS 1

$$\frac{1}{a_{\overline{n}|}} = a_{\overline{n}|}^{-1} = \frac{i}{1 - (1 + i)^{-n}} = \frac{i}{1 - v^n} = s_{\overline{n}|}^{-1} + i$$

Years n	Rate i				
	$0025(\frac{1}{4}\%)$	$.004167(\frac{5}{12}\%)$	$.005(\frac{1}{2}\%)$	$.005833(\frac{7}{12}\%)$	$.0075(\frac{3}{4}\%)$
1	1.00250000	1.00416667	1.00500000	1.00583333	1.00750000
2	0.50187578	0.50312717	0.50375312	0.50437924	0.50563200
3	.33500139	.33611496	.33667221	.33722976	.33834579
4	.25156445	.25260958	.25313279	.25365644	.25470501
5	.20150250	.20250693	.20300997	.20351357	.20452242
6	.16812803	.16910564	.16959546	.17008594	.17106891
7	.14428928	.14524800	.14572854	.14620986	.14717488
8	.12641035	.12735512	.12782886	.12830352	.12925552
9	.11250462	.11343876	.11390736	.11437698	.11531929
10	.10138015	.10230596	.10277057	.10323632	.10417123
11	.09227840	.09319757	.09365903	.09412175	.09505094
12	.08469370	.08560748	.08606643	.08652675	.08745148
13	.07827595	.07918532	.07964224	.08010064	.08102188
14	.07277510	.07368082	.07413609	.07459295	.07551146
15	.06800777	.06891045	.06936436	.06982000	.07073639
16	.06383642	.06473655	.06518937	.06564401	.06655879
17	.06015587	.06105387	.06150579	.06195966	.06287321
18	.05688433	.05778053	.05823173	.05868499	.05959766
19	.05395722	.05485191	.05530253	.05575532	.05666740
20	.05132288	.05221630	.05266645	.05311889	.05403063
21	.04893942	.04983183	.05028163	.05073383	.05164543
22	.04677278	.04766427	.04811380	.04856585	.04947748
23	.04479455	.04568531	.04613465	.04658663	.04749846
24	.04298121	.04387139	.04432061	.04477258	.04568474
25	.04131298	.04220270	.04265186	.04310388	.04401650
26	.03977316	.04066247	.04111163	.04156370	.04247693
27	.03834736	.03923645	.03968565	.04013793	.04105176
28	.03702347	.03791239	.03836167	.03881415	.03972871
29	.03579093	.03667974	.03712914	.03758186	.03849723
30	.03464059	.03552936	.03597892	.03643191	.03734816
31	.03356449	.03445330	.03490304	.03535633	.03627290
32	.03255569	.03344458	.03389453	.03434815	.03526634
33	.03160806	.03249708	.03294727	.03340124	.03432048
34	.03071620	.03160540	.03205586	.03251020	.03343053
35	.02987533	.03076476	.03121550	.03167024	.03259170
36	.02908121	.02997090	.03042194	.03087710	.03179973
37	.02833004	.02922003	.02967139	.03012698	.03105082
38	.02761843	.02850875	.02896045	.02941649	.03034157
39	.02694335	.02783402	.02828607	.02874258	.02966893
40	.02630204	.02719310	.02764552	.02810251	.02903016
41	.02569204	.02658352	.02703631	.02749379	.02842276
42	.02511112	.02600303	.02645622	.02691420	.02784452
43	.02455724	.02544961	.02590320	.02636170	.02729338
44	.02402855	.02492141	.02537541	.02583443	.02676751
45	.02352339	.02441675	.02487117	.02533073	.02626521
46	.02304022	.02393409	.02438894	.02484905	.02578495
47	.02257762	.02347204	.02392733	.02438798	.02532532
48	.02213433	.02302929	.02348503	.02394624	.02488504
49	.02170915	.02260468	.02306087	.02352265	.02446292
50	.02130099	.02219711	.02265376	.02311612	.02405787

ANNUITY WHOSE PRESENT VALUE IS 1

$$a_{\overline{n}|}^{-1} = i/(1 - v^n) = s_{\overline{n}|}^{-1} + i \text{ (Continued)}$$

Years n	Rate i				
	.0025($\frac{1}{4}$%)	.004167($\frac{5}{12}$%)	.005($\frac{1}{2}$%)	.005833($\frac{7}{12}$%)	.0075($\frac{3}{4}$%)
50	.02130099	.02219711	.02265376	.02311612	.02405787
51	.02090886	.02180557	.02226269	.02272563	.02366888
52	.02053184	.02142916	.02188675	.02235027	.02329503
53	.02016906	.02106700	.02152507	.02198919	.02293546
54	.01981974	.02071830	.02117686	.02164157	.02258938
55	.01948314	.02038234	.02084139	.02130671	.02225605
56	.01915858	.02005843	.02051797	.02098390	.02193478
57	.01884542	.01974593	.02020598	.02067251	.02162496
58	.01854308	.01944426	.01990481	.02037196	.02132597
59	.01825101	.01915287	.01961392	.02008170	.02103727
60	.01796869	.01887123	.01933280	.01980120	.02075836
61	.01769564	.01859888	.01906096	.01952999	.02048873
62	.01743142	.01833536	.01879796	.01926762	.02022795
63	.01717561	.01808025	.01854337	.01901360	.01997560
64	.01692780	.01783315	.01829681	.01876773	.01973127
65	.01668764	.01759371	.01805789	.01852946	.01949460
66	.01645476	.01736156	.01782627	.01829848	.01926524
67	.01622886	.01713639	.01760163	.01807449	.01904286
68	.01600961	.01691788	.01738366	.01785716	.01882716
69	.01579674	.01670574	.01717206	.01764622	.01861785
70	.01558996	.01649971	.01696657	.01744138	.01841464
71	.01538902	.01629952	.01676693	.01724239	.01821728
72	.01519368	.01610493	.01657289	.01704901	.01802554
73	.01500370	.01591572	.01638422	.01686100	.01783917
74	.01481887	.01573165	.01620070	.01667814	.01765796
75	.01463898	.01555253	.01602214	.01650024	.01748170
76	.01446385	.01537816	.01584832	.01632709	.01731020
77	.01429327	.01520836	.01567908	.01615851	.01714328
78	.01412708	.01504295	.01551423	.01599432	.01698074
79	.01396511	.01488177	.01535360	.01583436	.01682244
80	.01380721	.01472464	.01519704	.01567847	.01666821
81	.01365321	.01457144	.01504439	.01552650	.01651790
82	.01350298	.01442200	.01489552	.01537830	.01637136
83	.01335639	.01427620	.01475028	.01523373	.01622847
84	.01321330	.01413391	.01460855	.01509268	.01608908
85	.01307359	.01399500	.01447021	.01495501	.01595308
86	.01293714	.01385935	.01433513	.01482060	.01582034
87	.01280384	.01372685	.01420320	.01468935	.01569076
88	.01267357	.01359740	.01407431	.01456115	.01556423
89	.01254625	.01347088	.01394837	.01443588	.01544064
90	.01242177	.01334721	.01382527	.01431347	.01531989
91	.01230004	.01322629	.01370493	.01419380	.01520190
92	.01218096	.01310803	.01358724	.01407679	.01508657
93	.01206446	.01299234	.01347213	.01396236	.01497382
94	.01195044	.01287915	.01335950	.01385042	.01486356
95	.01183884	.01276836	.01324930	.01374090	.01475571
96	.01172957	.01265992	.01314143	.01363372	.01465020
97	.01162257	.01255374	.01303583	.01352880	.01454696
98	.01151776	.01244976	.01293242	.01342608	.01444592
99	.01141508	.01234790	.01283115	.01332549	.01434701
100	.01131446	.01224811	.01273194	.01322696	.01425017

INTEREST TABLES (Continued)

ANNUITY WHOSE PRESENT VALUE IS 1
$$a_{\overline{n}|}^{-1} = i/(1 - v^n) = s_{\overline{n}|}^{-1} + i \text{ (Continued)}$$

Years n	.01(1 %)	.01125(1⅛ %)	.0125(1¼ %)	.015(1½ %)	.0175(1¾ %)
			Rate i		
1	1.01000000	1.01125000	1.01250000	1.01500000	1.01750000
2	0.50751244	0.50845323	0.50939441	0.51127792	0.51316295
3	.34002211	.34086130	.34170117	.34338296	.34506746
4	.25628109	.25707058	.25786102	·.25944479	.26103237
5	.20603980	.20680034	.20756211	.20908932	.21062142
6	.17254837	.17329034	.17403381	.17552521	.17702256
7	.14862828	.14935762	.15008872	.15155616	.15303059
8	.13069029	.13141071	.13213314	.13358402	.13504292
9	.11674036	.11745432	.11817055	.11960982	.12105813
10	.10558208	.10629131	.10700307	.10843418	.10987534
11	.09645408	.09715984	.09786839	.09929384	.10073038
12	.08884879	.08955203	.09025831	.09167999	.09311377
13	.08241482	.08311626	.08382100	.08524036	.08667283
14	.07690117	.07760138	.07830515	.07972332	.08115562
15	.07212378	.07282321	.07352646	.07494436	.07637739
16	.06794460	.06864363	.06934672	.07076508	.07219958
17	.06425806	.06495698	.06566023	.06707966	.06851623
18	.06098205	.06168113	.06238479	.06380578	.06524492
19	.05805175	.05875120	.05945548	.06087847	.06232061
20	.05541531	.05611531	.05682039	.05824574	.05969122
21	.05303075	.05373145	.05443749	.05586550	.05731464
22	.05086372	.05156525	.05227238	.05370332	.05515638
23	.04888584	.04958833	.05029666	.05173075	.05318796
24	.04707347	.04777701	.04848665	.04992410	.05138565
25	.04540675	.04611144	.04682247	.04826345	.04972952
26	.04386888	.04457479	.04528729	.04673196	.04820269
27	.04244553	.04315273	.04386677	.04531527	.04679079
28	.04112444	.04183299	.04254863	.04400108	.04548151
29	.03989502	.04060498	.04132228	.04277878	.04426424
30	.03874811	.03945953	.04017854	.04163919	.04312975
31	.03767573	.03838866	.03910942	.04057430	.04207005
32	.03667089	.03738535	.03810791	.03957710	.04107812
33	.03572744	.03644349	.03716786	.03864144	.04014779
34	.03483997	.03555763	.03628387	.03776189	.03927363
35	.03400368	.03472299	.03545111	.03693363	.03845082
36	.03321431	.03393529	.03466533	.03615240	.03767507
37	.03246805	.03319072	.03392270	.03541437	.03694257
38	.03176150	.03248589	.03321983	.03471613	.03624990
39	.03109160	.03181773	.03255365	.03405463	.03559399
40	.03045560	.03118349	.03192141	.03342710	.03497209
41	.02985102	.03058069	.03132063	.03283106	.03438170
42	.02927563	.03000709	.03074906	.03226426	.03382057
43	.02872737	.02946064	.03020466	.03172465	.03328666
44	.02820441	.02893949	.02968557	.03121038	.03277810
45	.02770505	.02844197	.02919012	.03071976	.03229321
46	.02722775	.02796652	.02871675	.03025125	.03183043
47	.02677111	.02751173	.02826406	.02980342	.03138836
48	.02633384	.02707632	.02783075	.02937500	.03096569
49	.02591474	.02665910	.02741563	.02896478	.03056124
50	.02551273	.02625898	.02701763	.02857168	.03017391

$$a_{\overline{n}|}^{-1} = i/(1 - v^n) = s_{\overline{n}|}^{-1} + i \text{ (Continued)}$$

Years n	.01 (1 %)	.01125 (1⅛ %)	.0125 (1¼ %)	.015 (1½ %)	.0175 (1¾ %)
			Rate i		
50	.02551273	.02625898	.02701763	.02857168	.03017391
51	.02512680	.02587494	.02663571	.02819469	.02980269
52	.02475603	.02550606	.02626897	.02783287	.02944665
53	.02439956	.02515149	.02591653	.02748537	.02910492
54	.02405658	.02481043	.02557760	.02715138	.02877672
55	.02372637	.02448213	.02525145	.02683018	.02846129
56	.02340824	.02416592	.02493739	.02652106	.02815795
57	.02310156	.02386116	.02463478	.02622341	.02786606
58	.02280573	.02356726	.02434303	.02593661	.02758503
59	.02252020	.02328366	.02406158	.02566012	.02731430
60	.02224445	.02300985	.02378993	.02539343	.02705336
61	.02197800	.02274534	.02352758	.02513604	.02680172
62	.02172041	.02248969	.02327410	.02488751	.02655892
63	.02147125	.02224247	.02302904	.02464741	.02632455
64	.02123013	.02200329	.02279203	.02441534	.02609821
65	.02099667	.02177178	.02256268	.02419094	.02587952
66	.02077052	.02154758	.02234065	.02397386	.02566813
67	.02055136	.02133037	.02212560	.02376376	.02546372
68	.02033889	.02111985	.02191724	.02356033	.02526597
69	.02013280	.02091571	.02171527	.02336329	.02507459
70	.01993282	.02071769	.02151941	.02317235	.02488930
71	.01973870	.02052552	.02132941	.02298727	.02470985
72	.01955019	.02033896	.02114501	.02280779	.02453600
73	.01936706	.02015779	.02096600	.02263368	.02436750
74	.01918910	.01998177	.02079215	.02246473	.02420413
75	.01901609	.01981072	.02062325	.02230072	.02404570
76	.01884784	.01964442	.02045910	.02214146	.02389200
77	.01868416	.01948269	.02029953	.02198676	.02374285
78	.01852488	.01932536	.02014436	.02183645	.02359806
79	.01836983	.01917226	.01999341	.02169036	.02345748
80	.01821885	.01902323	.01984652	.02154832	.02332093
81	.01807179	.01887812	.01970356	.02141019	.02318828
82	.01792851	.01873678	.01956437	.02127583	.02305936
83	.01778887	.01859908	.01942881	.02114509	.02293406
84	.01765273	.01846489	.01929675	.02101784	.02281223
85	.01751998	.01833409	.01916808	.02089396	.02269375
86	.01739050	.01820654	.01904267	.02077333	.02257850
87	.01726418	.01808215	.01892041	.02065584	.02246636
88	.01714089	.01796081	.01880119	.02054138	.02235724
89	.01702056	.01784240	.01868491	.02042984	.02225102
90	.01690306	.01772684	.01857146	.02032113	.02214760
91	.01678832	.01761403	.01846076	.02021516	.02204690
92	.01667624	.01750387	.01835272	.02011182	.02194882
93	.01656673	.01739629	.01824724	.02001104	.02185327
94	.01645971	.01729119	.01814425	.01991273	.02176017
95	.01635511	.01718851	.01804366	.01981681	.02166944
96	.01625284	.01708816	.01794541	.01972321	.02158101
97	.01615284	.01699007	.01784941	.01963186	.02149480
98	.01605503	.01689418	.01775560	.01954268	.02141074
99	.01595936	.01680041	.01766391	.01945560	.02132876
100	.01586574	.01670870	.01757428	.01937057	.02124880

ANNUITY WHOSE PRESENT VALUE IS 1

$$a_{\overline{n}|}^{-1} = i/(1 - v^n) = s_{\overline{n}|}^{-1} + i \text{ (Continued)}$$

Years n	.02(2%)	.0225(2¼%)	.025(2½%)	.0275(2¾%)	.03(3%)
			Rate i		
1	1.02000000	1.02250000	1.02500000	1.02750000	1.03000000
2	0.51504950	0.51693758	0.51882716	0.52071825	0.52261084
3	.34675467	.34844458	.35013717	.35183243	.35353036
4	.26262375	.26421893	.26581788	.26742059	.26902705
5	.21215839	.21370021	.21524686	.21679832	.21835457
6	.17852581	.18003496	.18154997	.18307083	.18459750
7	.15451196	.15600025	.15749543	.15899747	.16050635
8	.13650980	.13798462	.13946735	.14095795	.14245639
9	.12251544	.12398170	.12545689	.12694095	.12843386
10	.11132653	.11278768	.11425876	.11573972	.11723051
11	.10217794	.10363649	.10510596	.10658629	.10807745
12	.09455960	.09601740	.09748713	.09896871	.10046209
13	.08811835	.08957686	.09104827	.09253252	.09402954
14	.08260197	.08406230	.08553652	.08702457	.08852634
15	.07782547	.07928852	.08076646	.08225917	.08376658
16	.07365013	.07511663	.07659899	.07809710	.07961085
17	.06996984	.07144039	.07292777	.07443186	.07595253
18	.06670210	.06817720	.06967008	.07118063	.07270870
19	.06378177	.06526182	.06676062	.06827802	.06981388
20	.06115672	.06264207	.06414713	.06567173	.06721571
21	.05878477	.06027572	.06178733	.06331941	.06487178
22	.05663140	.05812821	.05964661	.06118640	.06274739
23	.05466810	.05617097	.05769638	.05924410	.06081390
24	.05287110	.05438023	.05591282	.05746863	.05904742
25	.05122044	.05273599	.05427592	.05583997	.05742787
26	.04969923	.05122134	.05276875	.05434116	.05593329
27	.04829309	.04982188	.05137687	.05295776	.05456421
28	.04698967	.04852525	.05008793	.05167738	.05329323
29	.04577836	.04732081	.04889127	.05048935	.05211467
30	.04464992	.04619934	.04777764	.04938442	.05101926
31	.04359635	.04515280	.04673900	.04835453	.04999893
32	.04261061	.04417415	.04576831	.04739263	.04904662
33	.04168653	.04325722	.04485938	.04649253	.04815612
34	.04081867	.04239655	.04400675	.04564875	.04732196
35	.04000221	.04158731	.04320558	.04485645	.04653929
36	.03923285	.04082522	.04245158	.04411132	.04580379
37	.03850678	.04010643	.04174090	.04340953	.04511162
38	.03782057	.03942753	.04107012	.04274764	.04445934
39	.03717114	.03878543	.04043615	.04212256	.04384385
40	.03655575	.03817738	.03983623	.04153151	.04326238
41	.03597188	.03760087	.03926786	.04097200	.04271241
42	.03541729	.03705364	.03872876	.04044175	.04219167
43	.03488993	.03653364	.03821688	.03993871	.04169811
44	.03438794	.03603901	.03773037	.03946100	.04122985
45	.03390962	.03556805	.03726751	.03900693	.04078518
46	.03345342	.03511921	.03682676	.03857493	.04036254
47	.03301792	.03469107	.03640669	.03816358	.03996051
48	.03260184	.03428233	.03600599	.03777158	.03957777
49	.03220396	.03389179	.03562348	.03739773	.03921314
50	.03182321	.03351836	.03525806	.03704092	.03886549

INTEREST TABLES (Continued)

ANNUITY WHOSE PRESENT VALUE IS 1

$$a_{\overline{n}|}^{-1} = i/(1 - v^n) = s_{\overline{n}|}^{-1} + i \text{ (Continued)}$$

Years n	Rate i				
	.02 (2 %)	.0225 (2¼ %)	.025 (2½ %)	.0275 (2¾ %)	.03 (3 %)
50	.03182321	.03351836	.03525806	.03704092	.03886549
51	.03145856	.03316102	.03490870	.03670014	.03853382
52	.03110909	.03281884	.03457446	.03637444	.03821718
53	.03077392	.03249094	.03425449	.03606297	.03791471
54	.03045226	.03217654	.03394799	.03576491	.03762558
55	.03014337	.03187489	.03365419	.03547953	.03734907
56	.02984656	.03158530	.03337243	.03520612	.03708447
57	.02956120	.03130712	.03310204	.03494404	.03683114
58	.02928667	.03103977	.03284244	.03469270	.03658848
59	.02902243	.03078268	.03259307	.03445153	.03635593
60	.02876797	.03053533	.03235340	.03422002	.03613296
61	.02852278	.03029724	.03212294	.03399767	.03591908
62	.02828643	.03006795	.03190126	.03378402	.03571385
63	.02805848	.02984704	.03168790	.03357866	.03551682
64	.02783855	.02963411	.03148249	.03338118	.03532760
65	.02762624	.02942878	.03128463	.03319120	.03514581
66	.02742122	.02923070	.03109398	.03300837	.03497110
67	.02722316	.02903955	.03091021	.03283236	.03480313
68	.02703173	.02885500	.03073300	.03266285	.03464159
69	.02684665	.02867677	.03056206	.03249955	.03448618
70	.02666765	.02850458	.03039712	.03234218	.03433663
71	.02649446	.02833816	.03023790	.03219048	.03419266
72	.02632683	.02817728	.03008417	.03204420	.03405404
73	.02616454	.02802169	.02993568	.03190311	.03392053
74	.02600736	.02787118	.02979222	.03176698	.03379191
75	.02585508	.02772554	.02965358	.03163560	.03366796
76	.02570751	.02758457	.02951956	.03150878	.03354849
77	.02556447	.02744808	.02938997	.03138633	.03343331
78	.02542576	.02731589	.02926463	.03126806	.03332224
79	.02529123	.02718784	.02914338	.03115382	.03321510
80	.02516071	.02706376	.02902605	.03104342	.03311175
81	.02503405	.02694350	.02891248	.03093674	.03301201
82	.02491110	.02682692	.02880254	.03083361	.03291576
83	.02479173	.02671387	.02869608	.03073389	.03282284
84	.02467581	.02660423	.02859298	.03063747	.03273313
85	.02456321	.02649787	.02849310	.03054420	.03264650
86	.02445381	.02639467	.02839633	.03045397	.03256284
87	.02434750	.02629452	.02830255	.03036667	.03248202
88	.02424416	.02619730	.02821165	.03028219	.03240393
89	.02414370	.02610291	.02812353	.03020041	.03232848
90	.02404602	.02601126	.02803809	.03012125	.03225556
91	.02395101	.02592224	.02795523	.03004460	.03218508
92	.02385859	.02583577	.02787486	.02997038	.03211694
93	.02376868	.02575176	.02779690	.02989850	.03205107
94	.02368118	.02567012	.02772126	.02982887	.03198737
95	.02359602	.02559078	.02764786	.02976141	.03192577
96	.02351313	.02551366	.02757662	.02969605	.03186619
97	.02343242	.02543868	.02750747	.02963272	.03180856
98	.02335383	.02536578	.02744034	.02957134	.03175281
99	.02327729	.02529489	.02737517	.02951185	.03169886
100	.02320274	.02522594	.02731188	.02945418	.03164667

ANNUITY WHOSE PRESENT VALUE IS 1
$$a_{\overline{n}|}^{-1} = i/(1 - v^n) = s_{\overline{n}|}^{-1} + i \text{ (Continued)}$$

Years n	.035(3½%)	.04(4%)	.045(4½%)	.05(5%)	.055(5½%)
1	1.03500000	1.04000000	1.04500000	1.05000000	1.05500000
2	0.52640049	0.53019608	0.53399756	0.53780488	0.54161800
3	.35693418	.36034854	.36377336	.36720856	.37065407
4	.27225114	.27549005	.27874365	.28201183	.28529449
5	.22148137	.22462711	.22779164	.23097480	.23417644
6	.18766821	.19076190	.19387839	.19701747	.20017895
7	.16354449	.16660961	.16970147	.17281982	.17596442
8	.14547665	.14852783	.15160965	.15472181	.15786401
9	.13144601	.13449299	.13757447	.14069008	.14383946
10	.12024137	.12329094	.12637882	.12950457	.13266777
11	.11109197	.11414904	.11724818	.12038889	.12357065
12	.10348395	.10655217	.10966619	.11282541	.11602923
13	.09706157	.10014373	.10327535	.10645577	.10968426
14	.09157073	.09466897	.09782032	.10102397	.10427912
15	.08682507	.08994110	.09311381	.09634229	.09962560
16	.08268483	.08582000	.08901537	.09226991	.09558254
17	.07904313	.08219852	.08541758	.08869914	.09204197
18	.07581684	.07899333	.08223690	.08554622	.08891992
19	.07294033	.07613862	.07940734	.08274501	.08615006
20	.07036108	.07358175	.07687614	.08024259	.08367933
21	.06803659	.07128011	.07460057	.07799611	.08146478
22	.06593207	.06919381	.07254565	.07597051	.07947123
23	.06401880	.06730906	.07068249	.07413682	.07766965
24	.06227283	.06558683	.06898703	.07247090	.07603580
25	.06067404	.06401196	.06743903	.07095246	.07454935
26	.05920540	.06256738	.06602137	.06956432	.07319307
27	.05785241	.06123854	.06471904	.06829186	.07195228
28	.05660265	.06001298	.06352081	.06712253	.07081440
29	.05544538	.05887993	.06241461	.06604551	.06976857
30	.05437133	.05783010	.06139154	.06505144	.06880539
31	.05337240	.05685535	.06044345	.06413212	.06791665
32	.05244150	.05594859	.05956320	.06328042	.06700519
33	.05157242	.05510357	.05874453	.06249004	.06633469
34	.05075966	.05431477	.05798191	.06175545	.06562958
35	.04999835	.05357732	.05727045	.06107171	.06497493
36	.04928416	.05288688	.05660578	.06043446	.06436635
37	.04861325	.05223957	.05598402	.05983979	.06379993
38	.04798214	.05163192	.05540169	.05928423	.06327217
39	.04738775	.05106083	.05485567	.05876462	.06277991
40	.04682728	.05052349	.05434315	.05827816	.06232034
41	.04629822	.05001738	.05386158	.05782229	.06189090
42	.04579828	.04954020	.05340868	.05739471	.06148927
43	.04532539	.04908989	.05298235	.05699333	.06111337
44	.04487768	.04866454	.05258071	.05661625	.06076128
45	.04445343	.04826246	.05220202	.05626173	.06043127
46	.04405108	.04788205	.05184471	.05592820	.06012175
47	.04366919	.04752189	.05150734	.05561421	.05983129
48	.04330646	.04718065	.05118858	.05531843	.05955854
49	.04296167	.04685712	.05088722	.05503965	.05930230
50	.04263371	.04655020	.05060215	.05477674	.05906145

Reprinted with permission. From *Mathematical Tables from Handbook of Chemistry & Physics, Tenth Edition.* Copyright CRC Press, Boca Raton, Florida.

$$a_{\overline{n}|}^{-1} = i/(1 - v^n) = s_{\overline{n}|}^{-1} + i \text{ (Continued)}$$

Years	Rate i				
n	.06(6%)	.065(6½%)	.07(7%)	.075(7½%)	.08(8%)
1	1.06000000	1.06500000	1.07000000	1.07500000	1.08000000
2	0.54543689	0.54926150	0.55309179	0.55692771	0.56076923
3	.37410981	.37757570	.38105167	.38453763	.38803351
4	.28859149	.29190274	.29522812	.29856751	.30192080
5	.23739640	.24063454	.24389069	.24716472	.25045645
6	.20336263	.20656831	.20979580	.21304489	.21631539
7	.17913502	.18233137	.18555322	.18880032	.19207240
8	.16103594	.16423730	.16746776	.17072702	.17401476
9	.14702224	.15023803	.15348647	.15676716	.16007971
10	.13586796	.13910469	.14237750	.14568593	.14902949
11	.12679294	.13005521	.13335600	.13669747	.14007634
12	.11927703	.12256817	.12590199	.12927783	.13269502
13	.11296011	.11628256	.11965085	.12306420	.12652181
14	.10758491	.11094048	.11434494	.11779737	.12129685
15	.10296276	.10635278	.10979462	.11328724	.11682954
16	.09895214	.10237757	.10585765	.10939116	.11297687
17	.09544480	.09890633	.10242519	.10600003	.10962943
18	.09235654	.09585461	.09941260	.10302896	.10670210
19	.08962086	.09315575	.09675301	.10041090	.10412763
20	.08718456	.09075640	.09439293	.09809219	.10185221
21	.08500455	.08861333	.09228900	.09602937	.09983225
22	.08304557	.08669120	.09040577	.09418687	.09803207
23	.08127848	.08496078	.08871393	.09253528	.09642217
24	.07967900	.08339770	.08718902	.09105008	.09497796
25	.07822672	.08198148	.08581052	.08971067	.09367878
26	.07690435	.08069480	.08456103	.08849961	.09250713
27	.07569717	.07952288	.08342573	.08740204	.09144810
28	.07459255	.07845305	.08239193	.08640520	.09048891
29	.07357961	.07747440	.08144865	.08549811	.08961854
30	.07264891	.07657744	.08058640	.08467124	.08882743
31	.07179222	.07575393	.07979691	.08391628	.08810728
32	.07100234	.07499665	.07907292	.08322599	.08745081
33	.07027293	.07429924	.07840807	.08259397	.08685163
34	.06959843	.07365610	.07779674	.08201461	.08630411
35	.06897386	.07306226	.07723396	.08148291	.08580326
36	.06839483	.07251332	.07671531	.08099447	.08534467
37	.06785743	.07200534	.07623685	.08054533	.08492440
38	.06735812	.07153480	.07579505	.08013197	.08453894
39	.06689377	.07109854	.07538676	.07975124	.08418513
40	.06646154	.07069373	.07500914	.07940031	.08386016
41	.06605886	.07031779	.07465602	.07907663	.08356149
42	.06568342	.06996842	.07433591	.07877789	.08328684
43	.06533312	.06964352	.07403590	.07850201	.08303414
44	.06500606	.06934119	.07375769	.07824710	.08280152
45	.06470050	.06905968	.07349957	.07801146	.08258728
46	.06441485	.06879743	.07325996	.07779354	.08238991
47	.06414768	.06855300	.07303744	.07759190	.08220799
48	.06389765	.06832505	.07283070	.07740527	.08204027
49	.06366356	.06811240	.07263853	.07723247	.08188557
50	.06344429	.06791393	.07245985	.07707241	.08174286

CHAPTER 6

Bond Price Calculation

Bond price calculations are a special combination of annuity certain and present value. We will use the concepts and calculations developed in previous chapters to show how to compute a bond price. You will not find this terribly difficult to understand and to do. Then why do we have a special chapter on bond price calculations? Here are three reasons:

1. Bonds are especially important in finance. Bond offerings each year far exceed stock offerings in value. Bond issuers include national, state, and local governments and their authorities and agencies, foreign governments and international governmental organizations, and both domestic and foreign corporations. Only corporations can issue stocks. Many bonds have wide, active trading markets, and their ownership includes many individuals, governments and their agencies, and businesses. United States Treasury bonds, in particular, offer high quality, a wide choice of maturity dates, and excellent trading markets. These features make bonds particularly attractive for many investors. Many investors have made bonds, especially United States Treasuries, the backbone of their investment portfolio.

2. Bonds offer a basis for pricing many other securities and many derivative instruments. These trade in some relation to certain bonds; we say they "trade off" these bonds. Such securities include many mortgage-backed securities. Other securities may actually have their interest rates determined by the rates paid on certain particular bonds or on indices based on certain categories of bonds. Many derivative securities have their rates or prices determined by the rates or prices of high-quality and widely traded bonds, such as United States Treasury bonds.

3. Bond trading calculations that are used for actual money transactions have specific precision requirements, described in the applicable regulations. Answers must be calculated according to these regulations. Such answers are "right" answers. All other calculated answers are "wrong" answers. If you are responsible for these calculations, you should understand these requirements. Although they are not laws, these regulations usually have the force of law.

A book like this is not the place to find the specific requirements you will need. You should consult the latest regulations. If you work for a securities firm that has a compliance officer, he or she should be able to help you with this (it's actually part of this individual's job). In other cases, try to find a knowledgeable person to help you. If you have a friendly broker, perhaps your broker's compliance officer will help you. Most financial firms consider it exceedingly important to comply fully with all applicable regulations.

If you are at all active in almost any kind of financial analytic or investment activity, you should understand the bond form, how it works, and how its calculations are performed.

When you finish this chapter, you should understand how to compute a bond price and have some familiarity with and understanding of the equations for computing this price. You should understand the meanings of some terms used in the bond business, especially those connected with bond pricing calculations. You should know how bonds are described and how bond prices are expressed. You should understand how some bond features may affect bond prices, and you should have some understanding of how to estimate bond prices, and the effects of some bond features, without actually doing calculations. You should understand some of the regulatory requirements for some bond price calculations and how to get help in meeting these requirements. Finally, you should understand how to accrue discount and amortize premium.

WHAT IS A BOND?

A bond consists of a principal amount, usually approximately the amount borrowed, together with periodic interest payments until the principal amount is repaid, together with the last interest payment. The date on which the principal amount is repaid is called the maturity date of the bond, and the bond is said to mature on that date. The periodic interest payments are also called the coupon payments, or coupons. The income from the coupon payments is called the coupon income. The frequency of coupon payments usually depends on the country of issuance. In the United States, coupons are almost always paid semiannually. In most European countries, they are paid annually. Other countries may have other payment periods.

For example, the United States Treasury has a bond outstanding that matures on August 15, 2029. It carries a coupon rate of 6.5%. This bond, like almost all bonds issued in the United States, pays interest semiannually. It is issued in multiples of $1,000, with a minimum amount of $1,000.

This means that the bond will pay interest of $65 per year on a $1,000 bond. The interest will be paid in two installments, of $32.50 each, on February 15 and August 15, until August 15, 2029. On August 15, 2029, the last interest payment of $32.50 will be paid, along with the $1,000 of maturing principal. The bond will then be paid off.

For a $1,000 bond, the principal amount is $1,000, and the face (or par) amount is also $1,000. With multiples of $1,000, you could buy $17,000 face amount, but not $17,500.

HOW BONDS ARE DESCRIBED

The bond industry describes bonds in a standard form: the issuer name, coupon rate, maturity date or time to maturity, and the offering yield or price. The bonds we just discussed would be described as Treasury 6 1/2s, due 08/15/29. If they were offered at 105, or at a yield of 5.8, the description would be Treasury 6 1/2s, due 08/15/29, at a price of 105, or Treasury 6 1/2s, due 08/15/29, at a yield of 5.80, or possibly even Treasury 6 1/2s of '29 at a 5.80. The coupon rate and maturity date are part of the bond contract and cannot usually be changed, at least without some contractual procedure. The yield or price is a measure of the market and can change each moment, depending on market activity.

HOW TO READ A BOND MARKET REPORT

Before we start the equation for bond price calculation, you should know how bond prices are expressed and how to read a bond market report.

Most readers are probably familiar with stock price reporting. Stock prices are stated in dollars per share. Thus, if AT&T stock is quoted at 22, it means that it is quoted at $22 per share. One hundred shares will cost $2,200 at that price, plus whatever commission may be charged. Changes in price are reported in dollars and cents per share (or possibly fractions of a dollar, in some cases). For example, if AT&T is reported as down $\frac{1}{2}$ point, it means that the price of a share of AT&T fell $.50 per share at the close of trading from the close of trading on the previous trading day. This also means that the price of the last trade on that day was $.50 lower than the price of the last trade on the previous day.

Bond prices are quoted as a percentage of par, not dollars per bond. For example, if a bond is quoted at a price of 100, this means that the bond is quoted at 100% of par. Similarly, quoted prices of 97 and 105 mean 97% and 105% of par, respectively. If you sell $10,000 par value at a price of 97, you will receive $9,700 from the sale of your bonds, excluding any commissions you might pay, plus any accrued interest you might receive. If you buy $100,000 par (face) amount of bonds at 98.765, you will pay $98,765.00, plus accrued interest and whatever commissions might be required.

Bond quotations are frequently stated in yield. In the United States, the yield, divided by 2, is the same as the interest rate i, which we saw in the first few chapters. The stated yield (y) is a nominal annual rate, compounded semi-annually. The stated yield is expressed as y in the bond price calculation equations that follow. Remember that yield y, divided by 2, is also the internal rate of return, which sets the present value of the flow of the bond's interest and principal payments equal to the price of the bond.

WHAT IS A CALL FEATURE?

Many bonds have a contractual feature that allows the issuer to repay all or part of the bond issue before maturity. This feature is called a call feature, and the bond is said to be callable.

For example, the United States Treasury has a bond outstanding that has a 10% coupon and matures on May 15, 2010. It is callable at par (100) on May 15, 2005. This means that on May 15, 2005 (or any interest payment date thereafter), the United States Treasury may repay those bonds still outstanding. These bonds would be described as Treasury 10s, due 05/15/10, callable 05/15/05 at par.

Most mortgages in the United States have a call feature. The borrower may repay all or part of his or her mortgage, at any time, at face value, also called par.

Sometimes the repayment includes paying a small extra amount, called a call premium. The total amount paid is called the call price.

TABLE 6.1   Example of Call Features

Time to call from Dated date	Call price
Less than 15 years	Not callable
15 to 20 years	104
More than 20 to 25 years	103
More than 25 to 30 years	102
More than 30 to 35 years	101
More than 35 to 40 years	100

For example, in the case of the Treasury 10% bond mentioned previously, the call price is par (100) and the call premium is zero (0). If the XYZ Corporation, could call the bonds, but at a price of 103, the call price would be 103 and the call premium would be 3.

Some call features can get quite elaborate. Table 6.1 shows one from a widely held set of municipal issues, going back more than 40 years. The bonds can be called on any interest payment date.

At the time, the 15-year call protection made this a call feature favorable to the bondholder.

Some bond issues have a mandatory sinking fund, with a mandatory sinking fund call. A mandatory sinking fund is a series of payments designed to retire all, or almost all, of the bonds before maturity.

For example, suppose the XYZ Corporation bonds were issued in an amount of $10 million and matured in 20 years. Suppose the original bond contract required an annual principal payment of $1 million in 11 years, continuing annually thereafter. At the 20th year, only $1 million, would remain of the original issue; all the rest would have been paid off. The required annual payments of $1 million would be a mandatory sinking fund for the issue. Similarly, the monthly payments for most home mortgages in the United States include a principal payment, which reduces the outstanding mortgage balance each month, until the last payment pays off the relatively small balance still remaining.

Sometimes the mandatory sinking fund depends on the company's revenues. During the mid-1970s, some natural gas bonds had sinking funds whose payment size depended on the level of natural gas production.

So many types of call features exist that an investor looking at any callable bond should make sure to have a complete, accurate description of the call feature.

WHAT IS A PUT OPTION?

A put option is the opposite of a call option. The put option allows the owner of the bond, under certain conditions, to return the bond to the issuer and receive his or her money back; the owner is said to "put" the bond back to the issuer. This must be done under the terms of the original bond contract. Usually, this can only be done on certain dates, and usually the owner must give the issuer advance notice.

The most common bonds with a put feature are United States savings bonds. These allow the owner to return the bonds to the United States Treasury and receive back his or her original investment, plus accumulated interest. In early 2001, the owner had to hold the bonds at least 6 months to do this.

DISCOUNT SECURITIES

A fixed-income security that sells at a price below par is said to be selling at a discount. This discount is accrued, or accreted, over the life of the security. Later in this chapter, we will show how to accrue discount.

Sometimes the price is below par because the market as a whole has dropped or because that particular security's price has dropped. Such a discount is called "market discount." Special rules may apply to the taxation of accrual of market discount. These rules are beyond the scope of this book. You should consult your tax adviser on this question.

Sometimes a security may be originally offered at a price below par. In this case, the amount of that discount may be the "original issue discount" and once more, special rules may apply to the taxation of the accrual of original issue discount. Again, you should consult your tax adviser on this question. In the past, the Internal Revenue Service has considered accrual of original issue discount as interest income, to be reported and taxed as such.

Special rules apply to the taxation of income from Treasury bills. These are discussed in Chapter 9 on discount yield.

THE GENERAL EQUATION FOR COMPUTING A BOND PRICE, GIVEN THE YIELD

You can see from the previous description and examples that the usual bond consists of a series of coupon payments, together with a final maturity amount. The coupon payments are a set of equal payments, equally spaced in time. They form an annuity certain. We studied annuities certain in Chapter 5. We can evaluate the final maturity amount by using a present value calculation.

We studied present value calculations in Chapter 4. We therefore have all the ingredients needed to compute the bond price, given the yield. In doing this calculation and in all the bond price calculations in this book, we will use the American trade custom of semiannual bond coupon payments and yields stated as a nominal annual rate, compounded semiannually, unless stated otherwise. If your country has a different trade custom, you should make the appropriate changes in the equations.

We will look at the bond price calculation equations for municipal bonds, as set forth by the Municipal Securities Rulemaking Board (MSRB) in Rule G-33. The source for all the equations presented in this section is the Municipal Securities Rulemaking Board, although the equations in some cases may be slightly revised for teaching purposes. Calculations for other bonds will be similar. A close look at the municipal bond requirements provides an insight into the calculations required for most transactions in this important field of finance. We will also use the MSRB's notation in this section.

I present all the equations for price calculation because I think the presentation should be complete in a book of this sort. Other types of bonds will have similar requirements. Millions of dollars change hands each day based on these calculations. You should be fully aware of the requirements for these.

Computation of accrued interest, and yield given the bond's price, are discussed later in the book.

These equations look more fearsome than they actually are. Don't let the apparent complexity put you off. Study the equations and the meaning of the symbols. Then read the description of what the equation is about. Then return to the equation and study it some more. You should have no problem understanding what the equation says.

For bonds with periodic (semiannual) interest payments, and more than six months to maturity, the equation is

$$P = \left[\frac{RV}{\left(1+\frac{Y}{2}\right)^{N-1+\frac{E-A}{E}}} \right] + \left[\sum_{K=1}^{N} \frac{100 \bullet \frac{R}{2}}{\left(1+\frac{Y}{2}\right)^{K-1+\frac{E-A}{E}}} \right] - \left[100 \bullet \frac{A}{B} \bullet R \right] \qquad \text{Equation 6.1}$$

(Source: Municipal Securities Rulemaking Board)

P is the security price per $100 of par value. This is the same as a percentage of par, and it is what you want to compute. Remember that bond prices are expressed as a percentage of par, not as dollars per bond.

RV is the redemption value of the security per $100 of par value. For bonds maturing at par, RV will be par (100). For bonds being priced to the call, RV will be the call price. For example, if the bond is being priced to a call price of 102.5, then $RV = 102.5$. Other securities may have other redemption values, depending on the particular case.

R is the annual coupon (interest) rate, expressed as a decimal. For example, if the security has a coupon rate of 6%, $R = .06$. We divide R by 2 because coupons are paid semiannually.

Y is the yield, expressed as a decimal. For example, if the bond traded at a yield of 5%, $Y = .05$. We divide Y by 2 because the yields stated are nominal annual rates compounded semiannually.

N is the number of interest payments between the settlement date and the maturity (redemption) date, including the payment on the redemption date. N is expressed as a whole number. For example, if a bond settles on June 1, 2001, and matures on October 1, 2003, $N = 5$ (dates are 10/01/01, 04/01/02, 10/01/02, 04/01/03, and 10/01/03). If the bond is being priced to a call, N is the number of interest payments between the settlement date and the call date of the call feature you are pricing to.

A is the number of accrued days in the settlement period. In the example in the previous paragraph, for municipals, almost always $A = 60$ (04/01/01 to 06/01/01). We'll discuss day count conventions to compute accrued interest later in the book.

B is the number of days in the year. For most municipal transactions, $B = 360$. We'll discuss these conventions later in the book.

E is the number of days in the interest payment period in which the settlement day falls. For most municipal transactions, $E = 180$. Once more, we'll discuss these conventions later in the book.

A NOTE ON YIELD

As discussed before, you can see that the use of Y/2 in this chapter is the same as the use of the symbol i in the interest equations in chapters 1 through 5. For most bond transactions in the United States, yields are stated as nominal annual rates, compounded semiannually. Therefore, $Y/2 = i$, the i in the interest equations. Then why not use i in the first place and state the yields as semiannual rates? We use Y because the bond business uses the word "yield" and the symbol Y; we follow the trade custom of the bond business. For example, they might say, "The bond yields 5.42%." This does not mean that the bond yields 5.42% annually and that we use an annual rate of 5.42% to compute the bond price. It means that the bond pays 2.71% semiannually, and we compute the bond price using equations for the 2.71% yield and semiannual periods. If you are active in bonds, you will use the word "yield" and the symbol Y instead of interest rate and i. In this book, when we are speaking of bonds and presenting bond equations, we will use both symbols from time to time. There should be no confusion from this.

A NOTE ON ACCRUED DAYS IN THE SETTLEMENT PERIOD (A IN EQUATION 6.1) AND DATED DATE

The settlement period starts on the date of the last interest payment. But what happens if the upcoming interest payment is the first interest payment? This will happen in the case of new issues. The date used is the dated date of the bond. The dated date is, almost always, the date the bond starts to accrue interest.

Suppose a new issue of bonds is sold in mid-June 2002, has interest payments on June 1 and December 1, and matures on June 1, 2012. Its dated date would probably be June 1, 2002. Interest will start to accrue from June 1, 2002.

In a very few cases, the dated date might be some previous date, but interest starts to accrue on June 1, 2002. For example, the bond might be dated June 1, 2000, but have no coupons until December 1, 2002, with interest accrual starting June 1, 2002. This happened with some of the New Housing Authority bonds when they were issued. Although these were new issues, from the viewpoint of calculating accrued interest they used an assumed previous interest payment date, not the actual dated date of the bonds. The actual dated date might have been several years before the actual date of issue.

ANALYSIS OF THE EQUATION

Look at the first term. You see a maturity (or redemption) value divided by an interest rate term raised to a power. This is just a present value, so the first term is just the present value of the final maturity amount.

Look at the second term. This is the sum of a series of present values of the interest payments, so this term is just the present value of the annuity certain of the interest payments. We discussed both these concepts in earlier chapters. Note that this equation uses the actual sum of the present values of the individual coupon payments, not the equation for an annuity certain that we developed in Chapter 5.

Both terms have the $[1 + (Y/2)]$ term raised to a power that is not integral. Most bond trades have a settlement date that is not an interest payment date. The $[1 + (Y/2)]$ term is raised to a power computed from the actual number of days since the last interest payment. This is a form of day interpolation to calculate a price correct for the actual settlement day.

The third term is a new concept. It is the accrued interest on the bond, from the last interest payment date to the settlement date.

Almost all bonds in the United States trade on an "accrued interest" basis. This means that after the bond price has been agreed upon, the buyer pays to the seller an additional amount, called the "accrued interest." This is the amount of coupon interest earned on the bond since the last interest payment was made. The bond accrues coupon interest at an equal rate for each day counted in the interest payment period. In some conventions, including the convention for municipal bonds, not all days are counted.

Bonds in default are traded without any accrued interest. We say they are traded "flat." For example, when the Washington Public Power Supply System bonds defaulted, they were traded without accrued interest, or "flat."

Why is the accrued interest deducted from the first two terms to figure the bond price? The first two terms show the total present value of all the payments to be received in the future. They include the present value of the final maturity amount and the present value of the annuity certain of coupon payments. There aren't any other payments. The present value of these payments is the present value of the whole bond package. It is the total amount to be paid by the buyer to the seller in this transaction, and it includes any accrued interest amounts.

We could simply leave it at that, and I understand that is done in some countries. But in the United States, and most other countries, we have the trade custom of adding accrued interest to the agreed-upon price.

We know the total value, and we know the accrued interest to be added to the price. Therefore, to compute the price, we must deduct the accrued interest from the total value. The result is the price. The accrued interest will be added back to the price, as a separate item, in the trade confirmation statement to compute the total amount due from buyer to seller.

Understanding accrued interest is also important in understanding duration. We will discuss duration later in the book. The duration calculations include present values of all future payments. This includes accrued interest as well as the bond price. Of course, twice a year, accrued interest will equal zero.

Suppose the yield (Y) equals the coupon rate (R)? This actually occurs fairly frequently in new issues, because many investors like to pay prices of about par and the market responds to this investor preference. Years ago, for municipals, if the yield equaled the coupon rate, the price was simply set equal to par. This is no longer the case for municipals. The MSRB changed this trade custom, and now the price is computed using the standard formula. Of course, you can trade the bond at a dollar price of par, if you wish.

What difference does it make? Sometimes, it doesn't make enough difference to change the dollar price. Other times, though, it can make a greater difference. Here is an example.

Suppose you buy a 4% bond at a 4% yield, with the next coupon payment due in 3 months. Your accrued interest will be 1. In three months, you will receive interest of 2, and at a 4% yield your bond will be worth 100 (par).

Suppose you pay 100, plus 1 in accrued interest. Your total will be

Principal	100.00
Accrued interest	1.00
Total amount	101.00

You are entitled to earn 1% on this investment during the next 3 months, so you are entitled to earn 1.01. However your actual value will be 102, earning only 1.00, as follows:

Bond at 4%	100.00
Interest paid	2.00
Total amount	102.00

You should have

Bond at 4%	100.00
Accrued interest	1.00
Earnings @ 1%	1.01
Total	102.01, or .01 more

than you actually have. You have not earned the promised rate.

How can you adjust for this? Simply reduce the purchase price of the bond from 100.00 to 99.99. The purchase then looks like this:

Bond	99.99
Accrued interest	1.00
Total amount	100.99

You would still earn 1%, or 1.01 in 3 months. In 3 months, you would have

Bond	99.99
Accrued interest	1.00
Earnings @ 1%	1.01
Total amount	102.00, which is what you actually have.

(Results have been rounded to two decimal places, and compounding is quarterly, for teaching purposes.)

Note that for trades made close to an interest payment date, the results will be closer to 100, because there is little time to accrue interest from the previous payment or until the next payment.

This is one of the few cases where the MSRB changed a trade custom.

For bonds with periodic (semiannual) interest payments and 6 months or less to maturity or redemption:

$$P = \left[\frac{RV + \dfrac{100 \bullet R}{M}}{1 + \left(\dfrac{E - A}{E} \right) \bullet \dfrac{Y}{M}} \right] - \left[100 \bullet \frac{A}{B} \bullet R \right] \qquad \text{Equation 6.2}$$

(Source: Municipal Securities Rulemaking Board.)

In Equation 6.2, P, RV, R, E, A, B, R, and Y are as described they were previously and M is the number of interest payments per year standard for this security. For most municipal securities, $M = 2$.

You can see that in this case, we have simply the present value of the payment at maturity (the maturity amount plus the interest payment due at maturity), minus the accrued interest.

For securities paying interest only at redemption,

$$P = \left[\frac{RV + \left(100 \bullet \dfrac{DIR}{B} \bullet R \right)}{1 + \left(\dfrac{DIR - A}{B} \bullet Y \right)} \right] - \left[100 \bullet \frac{A}{B} \bullet R \right] \qquad \text{Equation 6.3}$$

(Source: Municipal Securities Rulemaking Board.)

In Equation 6.3, P, RV, R, A, B, R, and Y are as described they were previously and DIR = the number of days from the issue date to the redemption date, using the standard day count conventions. You can see that, in this case, the equation is similar to Equation 6.2. There is no compounding, regardless of the time from issue to maturity.

Theoretically, DIR can be any number of days. Usually, it is 1 year (360 days for municipals) or less. Securities priced using this equation usually are short-term securities.

STANDARDS OF ACCURACY

Here are the standards of accuracy for municipal security trades, as promulgated by the MSRB. Other securities will have similar standards.

Intermediate values are calculated to no fewer than 10 decimal places. Accrued interest is truncated to three decimal places, and the final result for the confirmation is rounded to two decimal places. Dollar prices are truncated to three decimal places. Yields are truncated to four decimal places and rounded to three decimal places, except that on most confirmations they may be accurate to the nearest .05% if the bond traded at a dollar price.

If you do computer or compliance work in the financial business, these regulations are very important to you. Most other persons need only to be aware of their existence.

PRICING ZERO COUPON BONDS

A zero coupon bond is a bond that has no coupons payable over the life of the bond. Instead, the bondholder receives the maturity amount on the maturity date. Usually, this maturity amount is 100, and the bond is sold at a discount, but sometimes the initial offering price is 100, and the bond matures for a larger amount. The initial price, or the final maturity amount, is set to give the bondholder the guaranteed return at maturity.

You can see that if $R = 0$, then the second and third terms in Equation 6.1 are zero, and the first term determines the price of the bond. This is simply the present value of the final maturity amount, at the agreed upon yield, as a nominal annual rate compounded semiannually.

The best-known zero coupon bonds are United States savings bonds. These have been offered since the 1930s and were heavily promoted during World War II. You can still buy them at your local bank or through a payroll deduction.

The accrual on zero coupon bonds, including United States savings bonds, may be subject to income tax, both federal and state, even though the investor receives no actual cash income. Consult with you tax accountant before you buy this kind of security.

PRICING TO A CALL FEATURE

If you compute the bond price using the call price as the redemption value and the call date as the maturity date, you have "priced the bond to the call." Usually, the buyer gets the bond priced to the call or to maturity, whichever is most favorable to the buyer (the lowest price), unless agreed to differently when the trade was made. If the bond has more than one call feature, like one of the previous examples, the price must be computed to each call feature and the most favorable to the buyer selected as the trade price.

Pricing to a put option is similar. You enter the put price and put date in the redemption value and time to maturity fields, and compute the price.

EXAMPLES OF BOND PRICE CALCULATIONS

Here are two examples of bond price calculations. We will compute them using the compound interest tables shown on pages 42–49 and 86–93. We will compute prices for a 4% bond due in 5 years at an 8% yield. We will then compute the price for a 12% bond due in 5 years at a 6% yield.

EXAMPLE 6.1. *A 4% bond, due in 5 years, at an 8% yield.* First of all, what does this phrase mean? The 4% is a statement about the bond, stating the coupon rate. This will be paid in semiannual payments of 2% each. The 5 years is also a statement about the bond, stating that it will mature in 5 years. Therefore, there will be 10 semiannual payments of 2%, starting in 6 months, and one final maturity payment of 100% of par in 5 years. These cannot be changed, except under the original bond contract. The 8% is a statement about the market and can change each moment in a fast-moving market. In this case, $y = .08$, so we use .04 as the actual interest rate for calculations. This is equivalent to $i = 4\%$ in our tables.

You can do a quick check for reasonableness of your answer when you compute a bond price or simply want to estimate the price without doing the calculation. Take the difference between the coupon and the yield, and multiply it by the number of years to maturity. Add the result for premium bonds, or subtract the result for discount bonds. Then adjust the result somewhat closer to par to allow for the present value calculations.

In this case, we have a 4% coupon, an 8% yield, and 5 years to maturity. The difference between coupon and yield is 4%. Multiplying this by 5 (years to maturity), we have 20. This gives us a price of about 80. Adjusting upward somewhat, we expect a price a little larger than 80. A price lower than 80, or larger than, say, 90 is probably wrong.

From the tables (or using your calculator), we have, at $i = 4\%$,

$$v^{10} = .67556417$$
$$a_{\overline{10}|} = 8.11089578$$

The price will be as follows:

The present value of the final maturity amount will be 100 (final maturity amount) times .67556417 = 67.556417

The present value of the coupon stream will be 2 (the size of the semiannual coupon payment) time 8.11089578 = 16.221792

Price of bond equals 67.556417 + 16.221792 = 83.778209
= 83.778, truncating to three places

This also agrees with our check for reasonableness, as somewhat larger than 80.

This bond is priced at a discount. When the coupon is less than the yield, the resulting price will be a discount (less than 100). The discount will be accrued (or accreted) over the life of the bond. We will discuss this process later.   ■

EXAMPLE 6.2.   *A 12% bond, due in 5 years, at a 6% yield.*   Here we have a coupon stream of 10 payments of size 6. Consulting the tables, or computing, we have

$$v^{10} = .74409391$$
$$a_{\overline{10}|} = 8.53020284$$

The present value of the final maturity amount will be 100 (par value) times .74409391 (the present value of 1) = 74.409391

The present value of the annuity will be 6 (the size of the semiannual coupon payment) times 8.53020284 (the present value of an annuity of 1) = 51.181217

Price of bond = 74.409391 + 51.181217 = 125.590608
125.590, truncating to three places

Note also that this agrees with the reasonableness check amount.

This bond is priced at a premium. If the coupon rate is higher than the yield, the bond will be priced at a premium.

The premium will be amortized over the life of the bond. We discuss this process in the next section.   ■

AMORTIZATION OF PREMIUM AND ACCRUAL OF DISCOUNT

Suppose you buy a bond at a discount or at a premium. You have paid a price different from par, but your bond will mature at par. This will give you a gain if you bought at a discount, or a loss if you bought at a premium. Usually, this premium or discount is written off or accrued over the life of the bond. Your actual earned income is adjusted each year to reflect this accrual or amortization. This can also have tax consequences, because it determines the book value, or cost basis, of your bond, as well as the interest income you report on your income tax returns. You should always consult with your tax adviser in doing this.

ACCRUAL OF DISCOUNT

When you buy a bond at a discount, you receive your return from two sources: the interest income you receive every 6 months and the gain you receive when

TABLE 6.2 Discount Accrual

Year	Earned	Interest paid	Increase in book value	Book value at end of period
0.0				83.778
0.5	3.351	2.000	1.351	85.129
1.0	3.405	2.000	1.405	86.534
1.5	3.461	2.000	1.461	87.996
2.0	3.520	2.000	1.520	89.515
2.5	3.581	2.000	1.581	91.096
3.0	3.644	2.000	1.644	92.740
3.5	3.710	2.000	1.710	94.450
4.0	3.778	2.000	1.778	96.228
4.5	3.849	2.000	1.849	98.077
5.0	3.923	2.000	1.923	100.000
Totals	36.222	20.000	16.222	

Calculations might not quite conform, due to rounding.

the bond matures. The interest is paid at 6 month intervals until maturity, but the gain is received only when the bond matures. For both accounting and portfolio management purposes, this gain is accrued over the time until maturity. Table 6.2 shows the discount accrual for the 4%, 5-year bond at an 8% yield.

You buy your bond at a price of 83.778. During the first half-year, you earn 4% of this amount (8% nominal annual rate, compounded semiannually), or 3.351. Part of this is your interest payment of 2.000. The rest, or 1.351, is added to the book value, which then becomes 85.129.

During the second half-year, you earn 4% of this new book value, or 3.405. Once more, 2.000 of this is your interest payment, and the rest, 1.405, is again added to the book value, which becomes 86.534.

After 4.5 years, the book value has become 98.077. In this last period, you earn 4% of this amount, or 3.923. Once more, the interest payment is 2.000, and the remainder of 1.923 is added to the book value. The book value becomes 100.000, and the bond matures.

You can see that the total earned is 36.222. Of this, 20.000 is interest paid over the life of the bond, and the rest, 16.222, is discount accrued (or accreted) over the life of the bond.

This is called the "yield basis" or "scientific" method of accruing discount. It is required for most income tax reporting, as well as accounting reports. Always consult with your tax adviser on matters like this.

You can see that this method accrues the discount in increasing amounts, as the book value increases. The earnings percentage rate stays the same, but the book value increases, so the actual earnings increase to keep up with it.

The book value is usually the cost basis of the bond. It is also the bond price at the original yield basis, so that if the market value of the bond stays at the same yield, the market price will follow the book values shown. Of course, this rarely happens in real markets.

Another way exists to accrue discount, called the straight-line method. This method accrues the discount in equal in equal installments over the life of the bond. In the previous example, discount would accrue in 10 installments of 1.6222 each. Formerly the straight-line method was frequently used, but it is rarely used today.

AMORTIZATION OF PREMIUM

Suppose you buy the 12% 5-year bond at a 6% yield and at a price of 125.590, as described earlier. Your premium is 25.590 points, and you amortize this, in semiannual installments, over the 5-year life of the bond. These amortization installments represent an actual return of a portion of your invested capital over the life of the bond. You should treat them as principal. Otherwise, you will have a loss of principal when the bond matures. Table 6.3. shows the premium amortization schedule for the bond in the previous example.

You buy your bond at a price of 125.590. During the first half-year, you earn 3% of this amount (6% nominal annual rate, compounded semiannually), or 3.768. You actually receive an interest payment of 6.000. Part of this is the 3.768 you actually earned. The rest, or 2.232, is returned principal and is subtracted from the book value, which then becomes 123.358.

TABLE 6.3   Premium Amortization Schedule

Year	Earned	Interest paid	Decrease in book value	Book value at end of period
0.0				125.590
0.5	3.768	6.000	2.232	123.358
1.0	3.701	6.000	2.299	121.059
1.5	3.632	6.000	2.368	118.691
2.0	3.561	6.000	2.439	116.252
2.5	3.488	6.000	2.512	113.740
3.0	3.412	6.000	2.588	111.152
3.5	3.335	6.000	2.665	108.487
4.0	3.255	6.000	2.745	105.742
4.5	3.172	6.000	2.828	102.914
5.0	3.087	6.000	2.913	100.000
Total	34.411	60.000	25.589	

Totals may not add due to rounding.

During the second half-year, you earn 3% of this new book value, or 3.701. Once more, you receive an interest payment of 6.000. Part of this is the 3.701 you actually earned. The rest, 2.299, is again subtracted from the book value, which becomes 121.059.

After 4.5 years, the book value has become 102.914. In this last period, you earn 3% of this amount, or 3.087. Once more, the interest payment is 6.000, and the difference of 2.913 is subtracted from the book value. The book value becomes 100.000, and the bond matures.

You can see that the total earned is 34.411. You received 60.000 of coupon payments over the life of the bond, but 25.589 of this amount was the premium you paid returned to you, and only the rest, 34.411, is actual earnings. The premium has been amortized over the life of the bond.

This is called the "yield basis" or "scientific" method of amortizing discount. It is required for most income tax reporting, as well as accounting reports. Always consult with your tax adviser on matters like this.

You can see that this method amortizes the premium in increasing amounts as the book value decreases. The earnings rate stays the same, but the book value decreases, so the earnings decrease as well. The book value is usually the cost basis of the bond. It is also the bond price at the original yield basis, so that if the market value of the bond stays at the same yield, the market price will follow the book values shown. Of course, this rarely happens in real markets.

Another way exists to amortize premium, called the straight-line method. This method amortizes the premium in equal installments over the life of the bond. In the previous example, the premium would be amortized in 10 installments of 2.559 each. The straight-line method was formerly frequently used, but is now used rarely.

A PORTFOLIO MANAGEMENT INTERLUDE

Suppose you are running a bond portfolio with annual maturities going out to 5 years or more. You know that, usually, longer-term bonds sell for higher yields than shorter-term bonds. However, you would like your shorter-term bonds to have the highest yield you can reasonably obtain. One way to do this is to buy longer-term bonds at a large premium. You amortize this premium over the life of the bond, just as we did in the previous paragraph. These premium amortizations may be considered as maturity amounts in the year they are amortized, because they are a return of the principal you invested when you bought the bond. You have made a shorter-term investment at a longer-term, and usually higher, rate.

For example, in the previous case, you bought a 12%, 5-year bond at a 6% yield. You paid 125.590, and in 6 months you will receive 2.232 of this back as premium to amortize. Suppose that a 6-month investment at the time you purchased the bond yielded 4.5%. You have made a 6-month investment of 2.232 at a 6% yield. Similarly, in 1 year, you will receive a 2.299 premium amortization amount. You have made a 1-year investment at a 5-year rate. Usually, this will be advantageous to you. Note that your premium bond must sell at a yield basis to make this an attractive purchase.

For example, in mid-September 2001, the Treasury 13.75s of 08/04 were trading at a yield of 3.40%. Treasuries maturing 08/02 were trading at yields of 2.42% to 2.47%, and Treasuries maturing 08/03 were trading at yields of 2.86 to 2.89%. Buying this 3-year Treasury at a high premium would have given you an extra 100 basis points of yield on a 1-year investment and 50 basis points on a 2-year investment. Remember, a basis point is one one-hundredth of a percent. These are yield increases of 1% and $\frac{1}{2}$% respectively, a large increase for such short-term investments.

Table 6.4 shows how such an investment might have worked out in mid-August 2001, with this bond.

In this case, you would have earned an extra 1% on the 4.757 returned principal over what you would have earned if you had bought a par bond maturing in 3 years and invested the 4.757 in a 1-year security.

PRICING A BOND TO A CALL

When you price a bond to the call feature, you use the call price as the redemption value and the time to the call date to compute the number of interest payments in the price calculation formula. Here is an example of pricing to a call.

TABLE 6.4   Amortizing the Premium of a 13.75% 3-Year Treasury Bond

Year	Earned	Interest paid	Decrease in book value	Book value at end of period
0.0				129.283
0.5	2.198	6.875	4.677	124.606
1.0	2.118	6.875	4.757	119.849
1.5	2.037	6.875	4.838	115.012
2.0	1.955	6.875	4.920	110.092
2.5	1.872	6.875	5.003	105.089
3.0	1.787	6.875	5.089	100.000
Total	11.967	41.250	29.283	

Totals may not add due to rounding.

CALL PRICE CALCULATION EXAMPLE

Suppose the 12%, 5-year bond priced at 6%, which we did previously, will be called in 5 years at a call price of 102. What will be the price?

Remember, we calculated the price as follows:

$$v^{10} = .74409391$$

$$a_{\overline{10}|} = 8.53020284$$

The present value of the final maturity amount was 100 (par value) times .74409391 (the present value of 1) = 74.409391

The present value of the annuity was 6 (the size of the semiannual coupon payment) times 8.53020284 (the present value of an annuity of 1) = 51.181217

$$\text{Price of bond} = 74.409391 = 51.181217 = 125.590608$$
$$= 125.590, \text{truncating to three places}$$

Now the present value of the final maturity amount will be 102 (redemption amount or call price) times .74409391, or 75.897579. The annuity stays the same, so its value stays the same. We then have

$$\text{Price of bond} = 75.897579 + 51.181217$$
$$= 127.078, \text{truncating to three places}$$

The price has been increased by the present value of the call premium. Note that in this calculation, the final maturity date is not stated and is not relevant to this calculation.

In computing the price for a customer confirmation, you should give the buyer what he or she considers to be the most advantageous price. For callable bonds, this means the bond is priced to the call or to maturity, whichever gives the lowest price. If the bond has several different call dates and prices, choose the call date and price that produces the lowest price.

For some callable bonds, general rules exist for pricing to call or to maturity. For bonds callable only at par, a yield lower than the coupon rate will result in pricing to the call. A yield higher than the coupon rate will result in pricing to maturity.

If the bond has a premium call, usually the individual price calculations must be made, and pricing to call or to maturity will depend on the results of those calculations. This could result in a number of calculations if the bond has several different premium calls.

However, a bond with a premium call could be priced at a premium, with the yield lower than the coupon, and still be priced to maturity, not to the call. Such pricings are said to "clear the call." This will happen when the yield is less than the coupon rate, but not low enough to make the price higher when the bond is priced to maturity than when it is priced to the call date and price.

Many customers do not want to buy bonds priced to the call, unless they pick up substantial additional yield. These bonds introduce additional complexity and more problems. Prices that clear the call avoid this problem. However, usually yields don't need to fall very much before such bonds will sell priced to the call. Some customers may not always be aware of this fact.

A LOOK AT A BASIS BOOK

Look at the tables on pages 129 to 130. This is an example from *Investors Bond Values Table*, a bond basis book published by Financial Publishing Company in Boston in 1962. A basis book contains prices of bond at various coupons, yields, and maturity dates. These pages show the prices for a 4% bond, due in 3 years to 3 years 11 months, at yields from 1% to 8.45%, in increments of .05%.

Basis books are hardly ever used now. Computers have replaced them for practically all uses. But they are still handy for looking at a variety of price/yield combinations quickly and easily, especially for prices (and yields) for one coupon maturing at one time or several closely related times. They can also be useful for teaching purposes.

These prices are shown to two decimal places. Therefore, they are not valid for use in municipal bond transactions and were not so valid in 1962 because these prices, now and in 1962, must be truncated to three decimal places. However, they are useful for other purposes.

Suppose you want to know the (approximate) price of a 4% bond, due in 3 years at a 2.5% yield. Look at the column for 3 years even (the leftmost price column), and go down the column until you find the price for a yield (in the leftmost column) of 2.50%. The price is 104.31.

Suppose you want to find the price of a 4% bond, due in 3 years and 4 months at a 2.75% yield. Look in the column for 3 years, 4 months, and go down the column until you find the price for a yield of 2.75%. The price is 103.95.

Note that for a 4% yield, the prices are all 100. This book was produced before the MSRB changed the rule (actually, before the MSRB even existed), so it followed the trade custom that the price equaled par if the yield equaled the coupon. As discussed earlier, this custom no longer holds.

BASIC RULES FOR PRICES AND YIELDS

Price moves inversely to yield. If you look down the columns for yields and for prices, you note that as yields increase, price decreases. You need to put aside less money to buy the promised flow of funds, because the money earns at a greater rate. This means that in a bull bond market, yields decrease because prices increase.

Increasing coupon increases price. If you increase coupon, you increase the total amount of money the owner of the bond receives. More money is always worth more money, so increasing coupon always increases price.

As time to maturity increases, premiums increase, and discounts increase. In the basis book diagram, look at the row across the page for a yield of 1.50%. You can see that the prices increases as you move across the page from left to right. If the bond trades at a premium, increasing the time to maturity increases the price. Now look at the row for a yield of 7.00%. You can see that the prices decrease as you move across the page. If the bond trades at a discount, increasing the time to maturity decreases the price. Par bond remain at par, at least on semiannual interest payment dates.

THE SHAPE OF THE PRICE-YIELD CURVE

Figure 6.1 shows the general shape of the price-yield curve. The x-axis shows yield, and the y-axis shows the corresponding price.

You can see that the price yield curve slopes downward, gradually becoming more horizontal but never quite actually becoming horizontal. As yield increases, it approaches zero (0) but never quite gets there. When yield equals

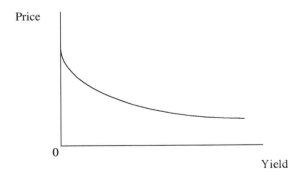

FIGURE 6.1   Price-Yield Curve

zero (0), the price becomes the total amount of the payments of interest and principal.

You can also see that the price-yield curve is convex. In Chapter 16 we'll discuss convexity and show how to measure it.

Suppose the bond has an embedded option, such as a call feature. Figure 6.2 shows the price of a bond that may be called now at a price of 100.

As the yield falls, the bond price continues to rise, until it reaches the call price of 100. At that point, even if the yield continues to decrease, the price will not increase because the bond can be called at 100, and no one will pay more.

What if there is only a chance that the bond will be called, and a chance it won't be? In that case, you adjust for the probabilities. We'll discuss probabilities in Chapter 18.

DIRTY PRICE AND CLEAN PRICE

Sometimes you will hear the terms "dirty price" and "clean price." These terms refer to whether or not the accrued interest is included in the number called "price." Dirty price includes accrued interest, so it is just the first two terms

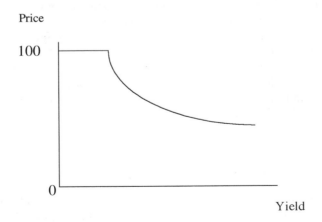

FIGURE 6.2   Price of a Bond

of Equation 6.1. Clean price does not include accrued interest, so it is the price computed according to Equation 6.1.

CHAPTER SUMMARY

Bond price calculations are especially important because (1) bonds are especially important in finance, (2) many other security prices are tied to bond prices, and (3) trade calculations have specific precision requirements.

A bond is defined as a principal amount combined with: periodic interest payments (every 6 months in the United States) until the final maturity date, when the principal is also repaid.

A call feature allows the issuer to repay all or part of the loan before maturity.

A put option allows the bond owner to demand repayment before maturity, according to the contract.

The bond price calculation is as follows:

$$P = \left[\frac{RV}{\left(1+\frac{Y}{2}\right)^{N-1+\frac{E-A}{E}}}\right] + \left[\sum_{K=1}^{N}\frac{100\bullet\frac{R}{2}}{\left(1+\frac{Y}{2}\right)^{K-1+\frac{E-A}{E}}}\right] - \left[100\bullet\frac{A}{B}\bullet R\right]$$

Zero coupon bond price calculations use only the first term.

Amortize premium and accrue discount by applying the yield earnings to the book value and adjust for interest payments.

When pricing to a call, use call price for redemption value and time to call date for the time to maturity.

Remember these general rules for prices and yields: (1) price moves inversely to yield, (2) increasing coupon increases price, and (3) increasing time to maturity increases prices for premium bonds and decreases prices for discount bonds.

COMPUTER PROJECTS

1. Plot the price of a 6% bond, maturing in 10 years, for yields from 0% to 20%. What happens to the price as the yield increases? What is the price for yield = 0? What is the shape of the price/yield curve? (This is

called convexity.) Now change the coupon, increasing it to 15% and decreasing it to 0%. What happens to the price/yield curve? Does the convexity increase or decrease as the coupon increases? Now change the time to maturity, decreasing it to 1 week, and increasing it to 30 years. What happens to the price yield curve? Does the convexity increase or decrease as the time to maturity increases?

2. You have just bought a 40-year, 5% bond, with the call features shown in Table 6.1. You bought it at a 5.05% yield. Will this bond be priced to any call features? Why, why not and, if so, which ones? Now lower the yield to 5%. Is the bond now priced to any call feature? If so, which ones? Continue to lower the yield by any reasonable, say .05, increments. At what point does the bond start being priced to a call? Which call is it priced to and why? Keep on lowering the yield. Does the bond become priced to any other call? Which one and why? As you keep lowering the yield, what calls show up in the pricing algorithm? At what yield to changes stop in the call feature used for pricing?

3. You have just bought a 40-year, 5% bond with the following call feature:
   • The bond is noncallable for 10 years.
   • The bond may be called on any interest payment date on or after 10 years from issue date.
   • The call premium will be $\frac{1}{2}$% times the number of full or partial years until final maturity.
   • The maximum call premium is 4%.

Go through the same procedure as in Computer Project 2. What results do you get? Do they differ from the results in Computer Project 2? Why or why not?

TOPICS FOR CLASS DISCUSSION

1. Some of the early zero coupon municipal bonds had a optional call at par (100) on or after 10 years from dated date. If you were the issuer and your investment banker proposed this to you, what would you comment?

2. You have just bought a 12-year zero coupon bond at a 6% yield to maturity. Your accountant has told you that you must pay a federal income tax, at 28%, on the accrual, even though you aren't actually collecting any cash. How will this affect the return on your investment? What will be the size of the effect, in dollars and in yield?

PROBLEMS

1. What are the three main terms in the bond price calculation equation, and what three main elements in the bond price do they describe?
2. What is the price of a 5% bond, due in 10 years, at a 4% yield?
3. A 5% bond, due in 30 years, and callable at par in 10 years, is priced at a 4% yield. Will it be priced to call or to maturity?
4. You have bought a 6%, 40-year bond at a 4% yield. The bond can be called in 10 years at 102. What are the dollar prices if the bond is (a) priced to maturity and (b) priced to the call? Should this bond be priced to the call or to maturity?

SUGGESTION FOR FURTHER STUDY

The following contains the rules for municipal bond transactions and can give an idea of how other rules will work: Municipal Securities Rulemaking Board, *MSRB Rule Book,* Municipal Securities Rulemaking Board, Alexandria, VA, 2001.

Investors Bond Values Table

4%   **3 YEARS**   **4%**

Yield	Even	1 mo	2 mo	3 mo	4 mo	5 mo	6 mo	7 mo	8 mo	9 mo	10 mo	11 mo
1.00	108.84	109.09	109.33	109.57	109.81	110.05	110.29	110.53	110.77	111.01	111.25	111.49
1.05	108.69	108.93	109.16	109.40	109.64	109.87	110.11	110.35	110.58	110.82	111.05	111.29
1.10	108.53	108.77	109.00	109.23	109.46	109.70	109.93	110.16	110.39	110.62	110.86	111.09
1.15	108.38	108.61	108.84	109.06	109.29	109.52	109.75	109.98	110.20	110.43	110.66	110.88
1.20	108.23	108.45	108.67	108.90	109.12	109.34	109.57	109.79	110.01	110.24	110.46	110.68
1.25	108.07	108.29	108.51	108.73	108.95	109.17	109.39	109.61	109.82	110.04	110.26	110.48
1.30	107.92	108.13	108.35	108.56	108.78	108.99	109.21	109.42	109.64	109.85	110.06	110.28
1.35	107.77	107.98	108.19	108.40	108.61	108.82	109.03	109.24	109.45	109.66	109.87	110.08
1.40	107.61	107.82	108.02	108.23	108.44	108.64	108.85	109.05	109.26	109.46	109.67	109.87
1.45	107.46	107.66	107.86	108.06	108.27	108.47	108.67	108.87	109.07	109.27	109.47	109.67
1.50	107.31	107.50	107.70	107.90	108.10	108.30	108.49	108.69	108.89	109.08	109.28	109.47
1.55	107.15	107.35	107.54	107.73	107.93	108.12	108.32	108.51	108.70	108.89	109.08	109.27
1.60	107.00	107.19	107.38	107.57	107.76	107.95	108.14	108.32	108.51	108.70	108.89	109.08
1.65	106.85	107.04	107.22	107.40	107.59	107.77	107.96	108.14	108.33	108.51	108.69	108.88
1.70	106.70	106.88	107.06	107.24	107.42	107.60	107.78	107.96	108.14	108.32	108.50	108.68
1.75	106.55	106.72	106.90	107.08	107.25	107.43	107.61	107.78	107.96	108.13	108.31	108.48
1.80	106.40	106.57	106.74	106.91	107.08	107.26	107.43	107.60	107.77	107.94	108.11	108.28
1.85	106.25	106.41	106.58	106.75	106.92	107.09	107.25	107.42	107.59	107.75	107.92	108.09
1.90	106.10	106.26	106.42	106.59	106.75	106.91	107.08	107.24	107.40	107.56	107.73	107.89
1.95	105.95	106.10	106.26	106.42	106.58	106.74	106.90	107.06	107.22	107.38	107.53	107.69
2.00	105.80	105.95	106.11	106.26	106.42	106.57	106.73	106.88	107.03	107.19	107.34	107.50
2.05	105.65	105.80	105.95	106.10	106.25	106.40	106.55	106.70	106.85	107.00	107.15	107.30
2.10	105.50	105.64	105.79	105.94	106.08	106.23	106.38	106.52	106.67	106.81	106.96	107.11
2.15	105.35	105.49	105.63	105.77	105.92	106.06	106.21	106.35	106.49	106.63	106.77	106.91
2.20	105.20	105.34	105.47	105.61	105.75	105.89	106.03	106.17	106.31	106.44	106.58	106.72
2.25	105.05	105.18	105.32	105.45	105.59	105.72	105.86	105.99	106.12	106.26	106.39	106.52
2.30	104.90	105.03	105.16	105.29	105.42	105.55	105.68	105.81	105.94	106.07	106.20	106.33
2.35	104.75	104.88	105.00	105.13	105.26	105.39	105.51	105.64	105.76	105.89	106.01	106.14
2.40	104.60	104.73	104.85	104.97	105.09	105.22	105.34	105.46	105.58	105.70	105.82	105.95
2.45	104.46	104.57	104.69	104.81	104.93	105.05	105.17	105.28	105.40	105.52	105.64	105.75
2.50	104.31	104.42	104.54	104.65	104.77	104.88	105.00	105.11	105.22	105.33	105.45	105.56
2.55	104.16	104.27	104.38	104.49	104.60	104.71	104.83	104.93	105.04	105.15	105.26	105.37
2.60	104.02	104.12	104.23	104.33	104.44	104.55	104.65	104.76	104.86	104.97	105.07	105.18
2.65	103.87	103.97	104.07	104.17	104.28	104.38	104.48	104.58	104.68	104.79	104.89	104.99
2.70	103.72	103.82	103.92	104.02	104.11	104.21	104.31	104.41	104.51	104.60	104.70	104.80
2.75	103.58	103.67	103.76	103.86	103.95	104.05	104.14	104.24	104.33	104.42	104.52	104.61
2.80	103.43	103.52	103.61	103.70	103.79	103.88	103.97	104.06	104.15	104.24	104.33	104.42
2.85	103.28	103.37	103.46	103.54	103.63	103.72	103.81	103.89	103.97	104.06	104.15	104.23
2.90	103.14	103.22	103.30	103.38	103.47	103.55	103.64	103.72	103.80	103.88	103.96	104.04
2.95	102.99	103.07	103.15	103.23	103.31	103.39	103.47	103.54	103.62	103.70	103.78	103.86
3.00	102.85	102.92	103.00	103.07	103.15	103.22	103.30	103.37	103.44	103.52	103.59	103.67
3.05	102.70	102.77	102.84	102.91	102.99	103.06	103.13	103.20	103.27	103.34	103.41	103.48
3.10	102.56	102.63	102.69	102.76	102.83	102.89	102.96	103.03	103.09	103.16	103.23	103.29
3.15	102.42	102.48	102.54	102.60	102.67	102.73	102.80	102.86	102.92	102.98	103.04	103.11
3.20	102.27	102.33	102.39	102.45	102.51	102.57	102.63	102.69	102.74	102.80	102.86	102.92
3.25	102.13	102.18	102.24	102.29	102.35	102.40	102.46	102.52	102.57	102.62	102.68	102.74
3.30	101.98	102.03	102.08	102.14	102.19	102.24	102.24	102.35	102.40	102.45	102.50	102.55
3.35	101.84	101.89	101.93	101.98	102.03	102.08	102.13	102.18	102.22	102.27	102.32	102.37
3.40	101.70	101.74	101.78	101.83	101.87	101.92	101.96	102.01	102.05	102.09	102.14	102.18
3.45	101.55	101.59	101.63	101.67	101.71	101.76	101.80	101.84	101.88	101.91	101.96	102.00
3.50	101.41	101.45	101.48	101.52	101.56	101.59	101.63	101.67	101.70	101.74	101.78	101.81
3.55	101.27	101.30	101.33	101.37	101.40	101.43	101.47	101.50	101.53	101.56	101.60	101.63
3.60	101.13	101.16	101.18	101.21	101.24	101.27	101.30	101.33	101.36	101.39	101.42	101.45
3.65	100.99	101.01	101.03	101.06	101.09	101.11	101.14	101.16	101.19	101.21	101.24	101.26
3.70	100.84	100.86	100.88	100.91	100.93	100.95	100.98	101.00	101.02	101.04	101.06	101.08
3.75	100.70	100.72	100.74	100.75	100.77	100.79	100.81	100.83	100.84	100.86	100.88	100.90
3.80	100.56	100.57	100.59	100.60	100.62	100.63	100.65	100.66	100.67	100.69	100.70	100.72
3.85	100.42	100.43	100.44	100.45	100.46	100.47	100.49	100.50	100.50	100.51	100.53	100.54
3.90	100.28	100.29	100.29	100.30	100.31	100.31	100.32	100.33	100.33	100.34	100.35	100.36
3.95	100.14	100.14	100.14	100.15	100.15	100.16	100.16	100.16	100.16	100.17	100.17	100.18
4.00	100.00	100.00	100.00	100.00	100.00	100.00	100.00	100.00	100.00	100.00	100.00	100.00
4.05	99.86	99.85	99.85	99.84	99.84	99.84	99.84	99.83	99.83	99.82	99.82	99.82
4.10	99.72	99.71	99.70	99.69	99.69	99.68	99.68	99.67	99.66	99.65	99.64	99.64
4.15	99.58	99.57	99.55	99.55	99.53	99.52	99.52	99.50	99.49	99.49	99.47	99.46
4.20	99.44	99.42	99.41	99.39	99.38	99.37	99.36	99.34	99.32	99.31	99.29	99.28
4.25	99.30	99.28	99.26	99.24	99.23	99.21	99.19	99.17	99.15	99.14	99.12	99.10
4.30	99.16	99.14	99.12	99.09	99.07	99.05	99.03	99.01	98.99	98.97	98.95	98.93
4.35	99.03	99.00	98.97	98.94	98.92	98.90	98.87	98.85	98.82	98.80	98.77	98.75
4.40	98.89	98.86	98.82	98.80	98.77	98.74	98.72	98.68	98.65	98.63	98.60	98.57
4.45	98.75	98.71	98.68	98.65	98.62	98.59	98.56	98.52	98.49	98.46	98.43	98.40
4.50	98.61	98.57	98.53	98.50	98.46	98.43	98.40	98.36	98.32	98.29	98.25	98.22
4.55	98.47	98.43	98.39	98.35	98.31	98.27	98.24	98.20	98.16	98.12	98.08	98.04
4.60	98.34	98.29	98.25	98.20	98.16	98.12	98.08	98.04	97.99	97.95	97.91	97.87
4.65	98.20	98.15	98.10	98.05	98.01	97.97	97.92	97.87	97.83	97.78	97.74	97.69
4.70	98.06	98.01	97.96	97.91	97.86	97.81	97.76	97.71	97.66	97.61	97.57	97.52

Reprinted with permission from Financial Publishing Company, South Bend, Indiana.

4% 3 YEARS 4%

Yield	Even	1 mo	2 mo	3 mo	4 mo	5 mo	6 mo	7 mo	8 mo	9 mo	10 mo	11 mo
4.75	97.93	97.87	97.81	97.76	97.71	97.66	97.61	97.55	97.50	97.45	97.39	97.34
4.80	97.79	97.73	97.67	97.61	97.56	97.50	97.45	97.39	97.33	97.28	97.22	97.17
4.85	97.65	97.59	97.53	97.47	97.41	97.35	97.29	97.23	97.17	97.11	97.05	97.00
4.90	97.52	97.45	97.38	97.32	97.26	97.20	97.14	97.07	97.01	96.94	96.88	96.82
4.95	97.38	97.31	97.24	97.17	97.11	97.04	96.98	96.91	96.84	96.78	96.71	96.65
5.00	97.25	97.17	97.10	97.03	96.96	96.89	96.83	96.75	96.68	96.61	96.55	96.48
5.05	97.11	97.03	96.96	96.88	96.81	96.74	96.67	96.59	96.52	96.45	96.38	96.31
5.10	96.98	96.89	96.81	96.74	96.66	96.59	96.51	96.44	96.36	96.28	96.21	96.14
5.15	96.84	96.76	96.67	96.59	96.51	96.44	96.36	96.28	96.20	96.12	96.04	95.96
5.20	96.71	96.62	96.53	96.45	96.36	96.28	96.20	96.12	96.03	95.95	95.87	95.79
5.25	96.57	96.48	96.39	96.30	96.22	96.13	96.05	95.96	95.87	95.79	95.70	95.62
5.30	96.44	96.34	96.25	96.16	96.07	95.98	95.90	95.80	95.71	95.62	95.54	95.45
5.35	96.30	96.21	96.11	96.01	95.92	95.83	95.74	95.65	95.55	95.46	95.37	95.28
5.40	96.17	96.07	95.97	95.87	95.78	95.68	95.59	95.49	95.39	95.30	95.20	95.11
5.45	96.04	95.93	95.83	95.73	95.63	95.53	95.44	95.33	95.23	95.13	95.04	94.94
5.50	95.90	95.79	95.69	95.58	95.48	95.38	95.28	95.18	95.07	94.97	94.87	94.77
5.55	95.77	95.66	95.55	95.44	95.34	95.23	95.13	95.02	94.91	94.81	94.71	94.61
5.60	95.64	95.52	95.41	95.30	95.19	95.08	94.98	94.87	94.76	94.65	94.54	94.44
5.65	95.50	95.39	95.27	95.16	95.04	94.93	94.83	94.71	94.60	94.49	94.38	94.27
5.70	95.37	95.25	95.13	95.01	94.90	94.79	94.67	94.56	94.44	94.33	94.21	94.10
5.75	95.24	95.12	94.99	94.87	94.75	94.64	94.52	94.40	94.28	94.16	94.05	93.94
5.80	95.11	94.98	94.85	94.73	94.61	94.49	94.37	94.25	94.12	94.00	93.89	93.77
5.85	94.98	94.85	94.72	94.59	94.46	94.34	94.22	94.09	93.97	93.84	93.72	93.60
5.90	94.85	94.71	94.58	94.45	94.32	94.19	94.07	93.94	93.81	93.68	93.56	93.44
5.95	94.71	94.58	94.44	94.31	94.18	94.05	93.92	93.79	93.65	93.52	93.40	93.27
6.00	94.58	94.44	94.30	94.17	94.03	93.90	93.77	93.63	93.50	93.36	93.23	93.11
6.05	94.45	94.31	94.17	94.03	93.89	93.75	93.62	93.48	93.34	93.21	93.07	92.94
6.10	94.32	94.17	94.03	93.89	93.74	93.61	93.47	93.33	93.19	93.05	92.91	92.78
6.15	94.19	94.04	93.89	93.75	93.60	93.46	93.32	93.17	93.03	92.89	92.75	92.61
6.20	94.06	93.91	93.75	93.61	93.46	93.31	93.17	93.02	92.88	92.73	92.59	92.45
6.25	93.93	93.77	93.62	93.47	93.32	93.17	93.02	92.87	92.72	92.57	92.43	92.28
6.30	93.80	93.64	93.48	93.33	93.17	93.02	92.88	92.72	92.57	92.42	92.27	92.12
6.35	93.67	93.51	93.35	93.19	93.03	92.88	92.73	92.57	92.41	92.26	92.11	91.96
6.40	93.54	93.38	93.21	93.05	92.89	92.73	92.58	92.42	92.26	92.10	91.95	91.80
6.45	93.41	93.24	93.08	92.91	92.75	92.59	92.43	92.27	92.11	91.95	91.79	91.63
6.50	93.28	93.11	92.94	92.77	92.61	92.44	92.29	92.12	91.95	91.79	91.63	91.47
6.55	93.16	92.98	92.81	92.63	92.47	92.30	92.14	91.97	91.80	91.63	91.47	91.31
6.60	93.03	92.85	92.67	92.50	92.33	92.16	91.99	91.82	91.65	91.48	91.31	91.15
6.65	92.90	92.72	92.54	92.36	92.19	92.01	91.85	91.67	91.49	91.32	91.15	90.99
6.70	92.77	92.59	92.40	92.22	92.05	91.87	91.70	91.52	91.34	91.17	91.00	90.83
6.75	92.64	92.45	92.27	92.09	91.91	91.73	91.55	91.37	91.19	91.01	90.84	90.67
6.80	92.52	92.32	92.13	91.95	91.77	91.59	91.41	91.22	91.04	90.86	90.68	90.51
6.85	92.39	92.19	92.00	91.81	91.63	91.44	91.26	91.07	90.89	90.71	90.53	90.35
6.90	92.26	92.06	91.87	91.68	91.49	91.30	91.12	90.93	90.74	90.55	90.37	90.19
6.95	92.13	91.93	91.73	91.54	91.35	91.16	90.97	90.78	90.59	90.40	90.21	90.03
7.00	92.01	91.80	91.60	91.40	91.21	91.02	90.83	90.63	90.44	90.25	90.06	89.87
7.05	91.88	91.67	91.47	91.27	91.07	90.88	90.68	90.48	90.29	90.09	89.90	89.71
7.10	91.75	91.54	91.34	91.13	90.93	90.73	90.54	90.34	90.14	89.94	89.75	89.56
7.15	91.63	91.41	91.20	91.00	90.79	90.59	90.40	90.19	89.99	89.79	89.59	89.40
7.20	91.50	91.29	91.07	90.86	90.66	90.45	90.25	90.04	89.84	89.64	89.44	89.24
7.25	91.38	91.16	90.94	90.73	90.52	90.31	90.11	89.89	89.69	89.48	89.28	89.08
7.30	91.25	91.03	90.81	90.59	90.38	90.17	89.97	89.75	89.54	89.33	89.13	88.93
7.35	91.13	90.90	90.68	90.46	90.25	90.03	89.82	89.61	89.39	89.18	88.98	88.77
7.40	91.00	90.77	90.55	90.33	90.11	89.89	89.68	89.46	89.25	89.03	88.82	88.61
7.45	90.88	90.64	90.42	90.19	89.97	89.75	89.54	89.32	89.10	88.88	88.67	88.46
7.50	90.75	90.52	90.29	90.06	89.84	89.62	89.40	89.17	88.95	88.73	88.52	88.30
7.55	90.63	90.39	90.16	89.93	89.70	89.48	89.26	89.03	88.80	88.58	88.36	88.15
7.60	90.50	90.26	90.03	89.79	89.56	89.34	89.12	88.88	88.66	88.43	88.21	87.99
7.65	90.38	90.14	89.90	89.66	89.43	89.20	88.98	88.74	88.51	88.28	88.06	87.84
7.70	90.25	90.01	89.77	89.53	89.29	89.06	88.83	88.60	88.36	88.13	87.91	87.69
7.75	90.13	89.88	89.64	89.40	89.16	88.92	88.69	88.45	88.22	87.99	87.76	87.53
7.80	90.01	89.76	89.51	89.26	89.02	88.79	88.55	88.31	88.07	87.84	87.61	87.38
7.85	89.88	89.63	89.38	89.13	88.89	88.65	88.41	88.17	87.93	87.69	87.46	87.23
7.90	89.76	89.50	89.25	89.00	88.75	88.51	88.27	88.03	87.78	87.54	87.31	87.07
7.95	89.64	89.38	89.12	88.87	88.62	88.38	88.13	87.88	87.64	87.40	87.16	86.92
8.00	89.52	89.25	88.99	88.74	88.49	88.24	88.00	87.74	87.49	87.25	87.01	86.77
8.05	89.39	89.13	88.87	88.61	88.35	88.10	87.86	87.60	87.35	87.10	86.86	86.62
8.10	89.27	89.00	88.74	88.48	88.22	87.97	87.72	87.46	87.21	86.96	86.71	86.47
8.15	89.15	88.88	88.61	88.35	88.09	87.83	87.58	87.32	87.06	86.81	86.56	86.31
8.20	89.03	88.75	88.48	88.22	87.95	87.70	87.44	87.18	86.92	86.66	86.41	86.16
8.25	88.91	88.63	88.36	88.09	87.82	87.56	87.30	87.04	86.78	86.52	86.26	86.01
8.30	88.78	88.50	88.23	87.96	87.69	87.43	87.17	86.90	86.63	86.37	86.12	85.86
8.35	88.66	88.38	88.10	87.83	87.56	87.29	87.03	86.76	86.49	86.23	85.97	85.71
8.40	88.54	88.26	87.98	87.70	87.43	87.16	86.89	86.62	86.35	86.08	85.82	85.56
8.45	88.42	88.13	87.85	87.57	87.29	87.02	86.76	86.48	86.21	85.94	85.67	85.41

CHAPTER 7

The Future Value (or Amount) of an Annuity

Suppose you make a periodic payment, of a fixed amount, into an investment fund, and the fund earns a given interest rate for a period of time. How much will it amount to at the end of the given time period? Your payments form an annuity certain and the value they will amount to are referred to as the future value of the annuity, or the amount of the annuity. When you finish this chapter, you should understand what the future value of an annuity means, how to compute its value, and where it might be used in practical work.

USES OF THE FUTURE VALUE OF AN ANNUITY

Here are some examples of the future value of an annuity.

1. You are in your employer's 401(k) program and contribute a certain amount every payday to your 401(k) fund. Your employer matches

part of your contribution. Assuming a given rate of return, how much will you have when you are ready to retire?

2. You are giving each of your children $12,000 annually, the maximum amount allowed without gift and estate tax considerations. How much will they have when you retire, assuming a given earnings rate in the funds?

3. You have bought an annuity from an insurance company and contribute a certain amount annually to the annuity. How much will you have at a given future time?

4. You contribute $2,000 annually to your IRA. How much will you have at retirement?

5. You contribute $500 monthly to a fund for the down payment on a house. When will you be able to buy a house?

6. You contribute $400 monthly to a fund for your children's college education. How much will you have when they go off to college?

Many readers will have such funds, especially as defined contribution retirement plans become more common and replace the old defined benefit plans, both as a corporate benefit and as an individual retirement effort. In early 2001, some people began suggesting that the Social Security system also have some self-managed plans as part of the Social Security program.

THE EQUATION FOR THE FUTURE VALUE OF AN ANNUITY

From Chapter 5, we have the equation for the present value of an annuity:

$$a_{\overline{n}} = \frac{\left(1 - \dfrac{1}{(1+i)^t}\right)}{i} \qquad \text{Equation 7.1}$$

To obtain the value t periods in the future, we simply multiply by $(1+i)^t$, compounding the value t periods into the future. We obtain

$$\text{Future value} = \frac{\left(1 - \dfrac{1}{(1+i)^t}\right)(1+i)^t}{i}$$
$$= \frac{\left[(1+i)^t - 1\right]}{i} \qquad \text{Equation 7.2}$$

This is the equation for the future value of an annuity of 1, after t periods, at interest rate i per period.

The extensions to continuous functions follow the outline in Chapter 3 and will not be repeated here.

THE TABLES FOR AMOUNT OF ANNUITY

Look at pages 135 to 142 for the amount (future value) of an annuity. You can see that as the interest rate i increases, the future value increases, because the amount deposited earns at a higher rate. As the number of periods increases, the amount increases, for two reasons. First, more payments have been made into the fund; second, the fund has been earning for a longer period.

SOME INVESTMENT POLICY IMPLICATIONS

You can see that as time passes, the fund increases. However, the contribution to the fund remains at 1. After a while, depending on the fund's earnings rate, the amount earned by the fund each period exceeds the amount contributed each period.

For example, after 13 years, at 6% interest, the value is 16.8699. The interest earnings during the next year will be 1.01, larger than the 1.00 contributed. At 19 years, the interest earnings are about double the annual contributions. You can see that after a relatively few years, at a reasonable interest rate, the management of the fund becomes more important than continuing the contributions, because the fund earns more than the contribution. From the viewpoint of individual financial management, this shows the importance of early and continued saving.

CHAPTER SUMMARY

The future value of an annuity is the amount accumulated, using both the periodic contributions and an assumed interest rate earned per period.

The amount equals the present value of an annuity multiplied by the amount at compound interest for the assumed interest rate and number of periods.

The equation for the future value of an annuity is

$$\frac{\left[(1+i)^t - 1\right]}{i}$$ for interest rate i and time t.

Practical applications include figuring accumulations in retirement and savings accounts.

COMPUTER PROJECT

You earn $30,000 annually and can contribute up to 10% into your 401(k) program, with your employer matching up to 4% of your salary. You contribute 5% of your salary annually. You wish to retire on $20,000 annually. Assume that you can earn 6% on your 401(k). When will you be able to retire? How will this change as the earnings rate changes? Try rates from 3% to 10% annually. Now change the contribution rate to 10% of earnings. How does this change your projected retirement date? Try earnings rates from 3% to 10% annually. Which is more important, increasing your contribution or increasing the earnings rate?

TOPIC FOR CLASS DISCUSSION

You wish to provide a comfortable retirement for yourself and your family. What factors should you consider in deciding on your periodic contribution to your retirement fund? How would you quantify these factors?

PROBLEMS

1. What is the relation between the equation for the present value of an annuity of 1 and the future value of an annuity of 1?
2. You contribute $1,000 each year, with the first contribution in 1 year, to a fund that earns 6.5% annually. How much will you have at the end of 17 years?
3. You plan to contribute $500 annually to a retirement plan for 10 years and then increase your contribution to $1,000 annually for another 20 years. You expect to earn 7% in the plan. How much will you have after 20 years?
4. You just got a raise and can now contribute $700 annually for 10 years, $1,200 annually for the next 5 years, and $1,500 annually for another 5 years. Now how much will you have in the plan after 20 years, if you expect to earn 6% annually?

AMOUNT OF ANNUITY $[(1 + i)^n - 1]/i$

The following table gives the amount of an annuity of unit value per period after a term of n periods at rate of interest of i per period; usually indicated as ($s_{\overline{n}|}$ at i).

n	.0025($\frac{1}{4}$%)	.004167($\frac{5}{12}$%)	.005($\frac{1}{2}$%)	.005833($\frac{7}{12}$%)	.0075($\frac{3}{4}$%)
1	1.00000000	1.00000000	1.00000000	1.00000000	1.00000000
2	2.00250000	2.00416667	2.00500000	2.00583333	2.00750000
3	3.00750625	3.01251736	3.01502500	3.01753403	3.02255625
4	4.01502502	4.02506952	4.03010013	4.03513631	4.04522542
5	5.02506258	5.04184064	5.05025063	5.05867460	5.07556461
6	6.03762523	6.06284831	6.07550188	6.08818354	6.11363135
7	7.05271930	7.08811018	7.10587939	7.12369794	7.15948358
8	8.07035110	8.11764397	8.14140879	8.16525285	8.21317971
9	9.09052697	9.15146749	9.18211583	9.21288349	9.27477856
10	10.1132533	10.1895986	10.2280264	10.2666253	10.3443394
11	11.1385364	11.2320553	11.2791665	11.3265140	11.4219219
12	12.1663828	12.2788555	12.3355624	12.3925853	12.5075864
13	13.1967987	13.3300174	13.3972402	13.4648754	13.6013933
14	14.2297907	14.3855591	14.4642264	14.5434205	14.7034037
15	15.2653652	15.4454990	15.5365475	15.6282571	15.8136792
16	16.3035286	16.5098552	16.6142303	16.7194219	16.9322818
17	17.3442874	17.5786463	17.6973014	17.8169519	18.0592739
18	18.3876481	18.6518906	18.7857879	18.9208841	19.1947185
19	19.4336173	19.7296068	19.8797169	20.0312559	20.3386789
20	20.4822013	20.8118135	20.9791154	21.1481049	21.4912190
21	21.5334068	21.8995294	22.0840110	22.2714689	22.6524031
22	22.5872403	22.9897733	23.1944311	23.4013858	23.8222961
23	23.6437084	24.0855640	24.3104032	24.5378939	25.0009634
24	24.7028177	25.1859205	25.4319552	25.6810316	26.1884706
25	25.7645747	26.2908619	26.5591150	26.8308376	27.3848841
26	26.8289862	27.4004071	27.6919106	27.9873508	28.5902707
27	27.8960587	28.5145755	28.8303701	29.1506104	29.8046978
28	28.9657988	29.6333862	29.9745220	30.3206556	31.0282330
29	30.0382133	30.7568587	31.1243946	31.4975261	32.2609448
30	31.1133088	31.8850122	32.2800166	32.6812616	33.5029018
31	32.1910921	33.0178665	33.4414167	33.8719023	34.7541736
32	33.2715698	34.1554409	34.6086237	35.0694884	36.0148299
33	34.3547488	35.2977552	35.7816669	36.2740604	37.2849411
34	35.4406356	36.4448292	36.9605752	37.4856591	38.5645782
35	36.5292372	37.5966827	38.1453781	38.7043255	39.8538125
36	37.6205603	38.7533355	39.3361050	39.9301007	41.1527161
37	38.7146117	39.9148078	40.5327855	41.1630263	42.4613615
38	39.8113982	41.0811195	41.7354494	42.4031440	43.7798217
39	40.9109267	42.2522908	42.9441267	43.6504956	45.1081704
40	42.0132041	43.4283420	44.1588473	44.9051235	46.4464816
41	43.1182041	44.6092934	45.3796415	46.1670701	47.7948303
42	44.2260327	45.7951655	46.6065397	47.4363780	49.1532915
43	45.3365977	46.9859787	47.8395724	48.7130902	50.5219412
44	46.4499392	48.1817536	49.0787703	49.9972499	51.9008557
45	47.5660641	49.3825109	50.3241642	51.2889005	53.2901121
46	48.6849792	50.5882713	51.5757850	52.5880858	54.6897880
47	49.8066917	51.7990558	52.8336639	53.8948496	56.0999614
48	50.9312084	53.0148852	54.0978322	55.2092362	57.5207111
49	52.0585364	54.2357806	55.3683214	56.5312901	58.9521164
50	53.1886828	55.4617630	56.6451630	57.8610559	60.3942573

AMOUNT OF ANNUITY $[(1 + i)^n - 1]/i$ (Continued)

Years n	.0025($\frac{1}{4}$ %)	.004167($\frac{5}{12}$ %)	.005($\frac{1}{2}$ %)	.005833($\frac{7}{12}$ %)	.0075($\frac{3}{4}$ %)
50	53.1886828	55.4617630	56.6451630	57.8610559	60.3942573
51	54.3216545	56.6928537	57.9283888	59.1985788	61.8472142
52	55.4574586	57.9290739	59.2180307	60.5439038	63.3110684
53	56.5961023	59.1704450	60.5141209	61.8970766	64.7859014
54	57.7375925	60.4169885	61.8166915	63.2581429	66.2717956
55	58.8819365	61.6687260	63.1257750	64.6271487	67.7688341
56	60.0291413	62.9256790	64.4414038	66.0041404	69.2771003
57	61.1792142	64.1878694	65.7636109	67.3891646	70.7966786
58	62.3321622	65.4553188	67.0924289	68.7822680	72.3276537
59	63.4879926	66.7280493	68.4278911	70.1834979	73.8701111
60	64.6467126	68.0060828	69.7700305	71.5929016	75.4241369
61	65.8083294	69.2894415	71.1188807	73.0105269	76.9898180
62	66.9728502	70.5781475	72.4744751	74.4364216	78.5672416
63	68.1402824	71.8722231	73.8368474	75.8706341	80.1564959
64	69.3106331	73.1716907	75.2060317	77.3132128	81.7576696
65	70.4839096	74.4765728	76.5820618	78.7642065	83.3708521
66	71.6601194	75.7868918	77.9649721	80.2236644	84.9961335
67	72.8392697	77.1026706	79.3547970	81.6916358	86.6336045
68	74.0213679	78.4239317	80.7515710	83.1681703	88.2833566
69	75.2064213	79.7506981	82.1553288	84.6533180	89.9454817
70	76.3944374	81.0829926	83.5661055	86.1471290	91.6200729
71	77.5854235	82.4208384	84.9839360	87.6496539	93.3072234
72	78.7793870	83.7642586	86.4088557	89.1609436	95.0070276
73	79.9763355	85.1132763	87.8409000	90.6810491	96.7195803
74	81.1762763	86.4679150	89.2801045	92.2100219	98.4449771
75	82.3792170	87.8281980	90.7265050	93.7479137	100.183314
76	83.5851651	89.1941488	92.1801375	95.2947765	101.934689
77	84.7941280	90.5657911	93.6410382	96.8506627	103.699199
78	86.0061133	91.9431485	95.1092434	98.4156249	105.476943
79	87.2211286	93.3262450	96.5847896	99.9897160	107.268021
80	88.4391814	94.7151044	98.0677136	101.572989	109.072531
81	89.6602793	96.1097506	99.5580521	103.165498	110.890575
82	90.8844300	97.5102079	101.055842	104.767297	112.722254
83	92.1116411	98.9165004	102.561122	106.378440	114.567671
84	93.3419202	100.328653	104.073927	107.998981	116.426928
85	94.5752750	101.746689	105.594297	109.628975	118.300130
86	95.8117132	103.170633	107.122268	111.268477	120.187381
87	97.0512425	104.600511	108.657880	112.917543	122.088787
88	98.2938706	106.036346	110.201169	114.576229	124.004453
89	99.5396053	107.478164	111.752175	116.244590	125.934486
90	100.788454	108.925990	113.310936	117.922684	127.878995
91	102.040425	110.379848	114.877490	119.610566	129.838087
92	103.295526	111.839764	116.451878	121.308294	131.811873
93	104.553765	113.305763	118.034137	123.015926	133.800462
94	105.815150	114.777871	119.624308	124.733519	135.803965
95	107.079688	116.256112	121.222430	126.461131	137.822495
96	108.347387	117.740512	122.828542	128.198821	139.856164
97	109.618255	119.231098	124.442684	129.946647	141.905085
98	110.892301	120.727894	126.064898	131.704670	143.969373
99	112.169532	122.230927	127.695222	133.472947	146.049143
100	113.449955	123.740222	129.333698	135.251539	148.144512

AMOUNT OF ANNUITY $[(1 + i)^n - 1]/i$ (Continued)

Years			Rate i		
n	.01(1 %)	.01125(1⅛ %)	.0125(1¼ %)	.015(1½ %)	.0175(1¾ %)
1	1.00000000	1.00000000	1.00000000	1.00000000	1.00000000
2	2.01000000	2.01125000	2.01250000	2.01500000	2.01750000
3	3.03010000	3.03387656	3.03765625	3.04522500	3.05280625
4	4.06040100	4.06800767	4.07562695	4.09090338	4.10623036
5	5.10100501	5.11377276	5.12657229	5.15226693	5.17808939
6	6.15201506	6.17130270	6.19065444	6.22955093	6.26870596
7	7.21353521	7.24072986	7.26803762	7.32299419	7.37840831
8	8.28567056	8.32218807	8.35888809	8.43283911	8.50753045
9	9.36852727	9.41581269	9.46337420	9.55933169	9.65641224
10	10.4622125	10.5217406	10.5816664	10.7027217	10.8253995
11	11.5668347	11.6401102	11.7139372	11.8632625	12.0148439
12	12.6825030	12.7710614	12.8603614	13.0412114	13.2251037
13	13.8093280	13.9147358	14.0211159	14.2368296	14.4565430
14	14.9474213	15.0712766	15.1963799	15.4503820	15.7095325
15	16.0968955	16.2408285	16.3863346	16.6821378	16.9844493
16	17.2578645	17.4235378	17.5911638	17.9323698	18.2816772
17	18.4304431	18.6195526	18.8110534	19.2013554	19.6016066
18	19.6147476	19.8290226	20.0461915	20.4893757	20.9446347
19	20.8108950	21.0520991	21.2967689	21.7967164	22.3111658
20	22.0190040	22.2889352	22.5629785	23.1236671	23.7016112
21	23.2391940	23.5396857	23.8450158	24.4705221	25.1163894
22	24.4715860	24.8045072	25.1430785	25.8375799	26.5559262
23	25.7163018	26.0835579	26.4573669	27.2251436	28.0206549
24	26.9734649	27.3769979	27.7880840	28.6335208	29.5110164
25	28.2431995	28.6849891	29.1354351	30.0630236	31.0274592
26	29.5256315	30.0076953	30.4996280	31.5139690	32.5704397
27	30.8208878	31.3452818	31.8808734	32.9866785	34.1404224
28	32.1290967	32.6979162	33.2793843	34.4814787	35.7378798
29	33.4503877	34.0657678	34.6953766	35.9987009	37.3632927
30	34.7848915	35.4490077	36.1290688	37.5386814	39.0171503
31	36.1327404	36.8478090	37.5806822	39.1017616	40.6999504
32	37.4940679	38.2623469	39.0504407	40.6882880	42.4121996
33	38.8690085	39.6927983	40.5385712	42.2986123	44.1544130
34	40.2576986	41.1393423	42.0453033	43.9330915	45.9271153
35	41.6602756	42.6021599	43.5708696	45.5920879	47.7308398
36	43.0768784	44.0814342	45.1155055	47.2759692	49.5661295
37	44.5076471	45.5773503	46.6794493	48.9851087	51.4335368
38	45.9527236	47.0900955	48.2629424	50.7198854	53.3336236
39	47.4122508	48.6198591	49.8662292	52.4806837	55.2669621
40	48.8863734	50.1668325	51.4895571	54.2678939	57.2341339
41	50.3752371	51.7312093	53.1331705	56.0819123	59.2357312
42	51.8789895	53.3131854	54.7973412	57.9231410	61.2723565
43	53.3977794	54.9129588	56.4823080	59.7919881	63.3446228
44	54.9317572	56.5307296	58.1883369	61.6888679	65.4531537
45	56.4810747	58.1667003	59.9156911	63.6142010	67.5985839
46	58.0458855	59.8210757	61.6646372	65.5684140	69.7815591
47	59.6263443	61.4940628	63.4354452	67.5519402	72.0027364
48	61.2226078	63.1858710	65.2283882	69.5652193	74.2627843
49	62.8348338	64.8967120	67.0437431	71.6086976	76.5623830
50	64.4631822	66.6268000	68.8817899	73.6828280	78.9022247

AMOUNT OF ANNUITY $[(1 + i)^n - 1]/i$ (Continued)

n (Years)	.01(1 %)	.01125(1⅛ %)	.0125(1¼ %)	.015(1½ %)	.0175(1¾ %)
50	64.4631822	66.6268000	68.8817899	73.6828280	78.9022247
51	66.1078140	68.3763515	70.7428123	75.7880705	81.2830136
52	67.7688921	70.1455855	72.6270974	77.9248915	83.7054663
53	69.4465811	71.9347233	74.5349361	80.0937649	86.1703120
54	71.1410469	73.7439890	76.4666228	82.2951714	88.6782925
55	72.8524573	75.5736088	78.4224556	84.5295989	91.2301626
56	74.5809819	77.4238119	80.4027363	86.7975429	93.8266904
57	76.3267917	79.2948298	82.4077705	89.0995061	96.4686575
58	78.0900597	81.1868966	84.4378676	91.4359987	99.1568590
59	79.8709603	83.1002492	86.4933410	93.8075386	101.892104
60	81.6696699	85.0351270	88.5745078	96.2146517	104.675216
61	83.4863666	86.9917722	90.6816891	98.6578715	107.507032
62	85.3212302	88.9704297	92.8152102	101.137740	110.388405
63	87.1744425	90.9713470	94.9754003	103.654806	113.320202
64	89.0461869	92.9947746	97.1625928	106.209628	116.303306
65	90.9366488	95.0409659	99.3771253	108.802772	119.338614
66	92.8460153	97.1101767	101.619339	111.434814	122.427039
67	94.7744755	99.2026662	103.889581	114.106336	125.569513
68	96.7222202	101.318696	106.188201	116.817931	128.766979
69	98.6894424	103.458532	108.515553	119.570200	132.020401
70	100.676337	105.622440	110.871998	122.363753	135.330758
71	102.683100	107.810692	113.257898	125.199209	138.699047
72	104.709931	110.023563	115.673621	128.077197	142.126280
73	106.757031	112.261328	118.119542	130.998355	145.613490
74	108.824601	114.524268	120.596036	133.963331	149.161726
75	110.912847	116.812666	123.103486	136.972781	152.772056
76	113.021975	119.126808	125.642280	140.027372	156.445567
77	115.152195	121.466985	128.212809	143.127783	160.183364
78	117.303717	123.833488	130.815469	146.274700	163.986573
79	119.476754	126.226615	133.450662	149.468820	167.856338
80	121.671522	128.646665	136.118795	152.710852	171.793824
81	123.888237	131.093940	138.820280	156.001515	175.800216
82	126.127119	133.568746	141.555534	159.341538	179.876720
83	128.388390	136.071395	144.324978	162.731661	184.024563
84	130.672274	138.602198	147.129040	166.172636	188.244992
85	132.978997	141.161473	149.968153	169.665226	192.539280
86	135.308787	143.749539	152.842755	173.210204	196.908717
87	137.661875	146.366722	155.753289	176.808357	201.354620
88	140.038494	149.013347	158.700206	180.460482	205.878326
89	142.438879	151.689747	161.683958	184.167390	210.481196
90	144.863267	154.396257	164.705008	187.929900	215.164617
91	147.311900	157.133215	167.763820	191.748849	219.929998
92	149.785019	159.900964	170.860868	195.625082	224.778773
93	152.282869	162.699849	173.996629	199.559458	229.712401
94	154.805698	165.530223	177.171587	203.552850	234.732369
95	157.353755	168.392438	180.386232	207.606142	239.840185
96	159.927293	171.286853	183.641059	211.720235	245.037388
97	162.526565	174.213830	186.936573	215.896038	250.325542
98	165.151831	177.173735	190.273280	220.134479	255.706239
99	167.803349	180.166940	193.651696	224.436496	261.181099
100	170.481383	183.193818	197.072342	228.803043	266.751768

AMOUNT OF ANNUITY $[(1 + i)^n - 1]/i$ (Continued)

n	.02(2 %)	.0225(2¼ %)	.025(2½ %)	.0275(2¾ %)	.03(3 %)
			Rate i		
1	1.00000000	1.00000000	1.00000000	1.00000000	1.00000000
2	2.02000000	2.02250000	2.02500000	2.02750000	2.03000000
3	3.06040000	3.06800625	3.07562500	3.08325625	3.09090000
4	4.12160800	4.13703639	4.15251563	4.16804580	4.18362700
5	5.20404016	5.23011971	5.25632852	5.28266706	5.30913581
6	6.30812096	6.34779740	6.38773673	6.42794040	6.46840988
7	7.43428338	7.49062284	7.54743015	7.60470876	7.66246218
8	8.58296905	8.65916186	8.73611590	8.81383825	8.89233605
9	9.75462843	9.85399300	9.95451880	10.0562188	10.1591061
10	10.9497210	11.0757078	11.2033818	11.3327648	11.4638793
11	12.1687154	12.3249113	12.4834663	12.6444159	12.8077957
12	13.4120897	13.6022218	13.7955530	13.9921373	14.1920296
13	14.6803315	14.9082718	15.1404448	15.3769211	15.6177904
14	15.9739382	16.2437079	16.5189528	16.7997864	17.0863242
15	17.2934169	17.6091913	17.9319267	18.2617805	18.5989139
16	18.6392853	19.0053981	19.3802248	19.7639795	20.1568813
17	20.0120710	20.4330196	20.8647304	21.3074889	21.7615877
18	21.4123124	21.8927625	22.3863487	22.8934449	23.4144354
19	22.8405586	23.3853497	23.9460074	24.5230146	25.1168684
20	24.2973698	24.9115200	25.5446576	26.1973975	26.8703745
21	25.7833172	26.4720292	27.1832741	27.9178259	28.6764857
22	27.2989835	28.0676499	28.8628559	29.6855661	30.5367803
23	28.8449632	29.6991720	30.5844273	31.5019192	32.4528837
24	30.4218625	31.3674034	32.3490380	33.3682220	34.4264702
25	32.0302997	33.0731700	34.1577639	35.2858481	36.4592643
26	33.6709057	34.8173163	36.0117080	37.2562089	38.5530423
27	35.3443238	36.6007059	37.9120007	39.2807547	40.7096335
28	37.0512103	38.4242218	39.8598008	41.3609754	42.9309225
29	38.7922345	40.2887668	41.8562958	43.4984022	45.2188502
30	40.5680792	42.1952640	43.9027032	45.6946083	47.5754157
31	42.3794408	44.1446575	46.0002707	47.9512100	50.0026782
32	44.2270296	46.1379123	48.1502775	50.2698683	52.5027585
33	46.1115702	48.1760153	50.3540344	52.6522897	55.0778413
34	48.0338016	50.2599756	52.6128853	55.1002277	57.7301765
35	49.9944776	52.3908251	54.9282074	57.6154839	60.4620818
36	51.9943672	54.5696186	57.3014126	60.1999097	63.2759443
37	54.0342545	56.7974351	59.7339479	62.8554072	66.1742226
38	56.1149396	59.0753774	62.2272966	65.5839309	69.1594493
39	58.2372384	61.4045733	64.7829791	68.3874890	72.2342328
40	60.4019832	63.7861762	67.4025535	71.2681450	75.4012597
41	62.6100228	66.2213652	70.0876174	74.2280190	78.6632975
42	64.8622233	68.7113459	72.8398078	77.2692895	82.0231965
43	67.1594678	71.2573512	75.6608030	80.3941950	85.4838923
44	69.5026571	73.8606416	78.5523231	83.6050353	89.0484091
45	71.8927103	76.5225060	81.5161312	86.9041738	92.7198614
46	74.3305645	79.2442624	84.5540344	90.2940386	96.5014572
47	76.8171758	82.0272583	87.6678853	93.7771246	100.396501
48	79.3535193	84.8728716	90.8595824	97.3559956	104.408396
49	81.9405897	87.7825113	94.1310720	101.033285	108.540648
50	84.5794015	90.7576178	97.4843488	104.811701	112.796867

AMOUNT OF ANNUITY $[(1 + i)^n - 1]/i$ (Continued)

Years n	.02(2 %)	.0225(2¼ %)	.025(2½ %)	.0275(2¾ %)	03(3 %)
50	84.5794015	90.7576178	97.4843488	104.811701	112.796867
51	87.2709895	93.7996642	100.921458	108.694023	117.180773
52	90.0164093	96.9101566	104.444494	112.683108	121.696197
53	92.8167375	100.090635	108.055606	116.781894	126.347082
54	95.6730722	103.342674	111.756996	120.993396	131.137495
55	98.5865337	106.667885	115.550921	125.320714	136.071620
56	101.558264	110.067912	119.439694	129.767034	141.153768
57	104.589430	113.544440	123.425687	134.335627	146.388381
58	107.681218	117.099190	127.511329	139.029857	151.780033
59	110.834843	120.733922	131.699112	143.853178	157.333434
60	114.051539	124.450435	135.991590	148.809140	163.053437
61	117.332570	128.250570	140.391380	153.901392	168.945040
62	120.679222	132.136208	144.901164	159.133680	175.013391
63	124.092806	136.109272	149.523693	164.509856	181.263793
64	127.574062	140.171731	154.261786	170.033877	187.701707
65	131.126155	144.325595	159.118330	175.709809	194.332758
66	134.748679	148.572921	164.096289	181.541829	201.162741
67	138.443652	152.915811	169.198696	187.534229	208.197623
68	142.212525	157.356417	174.428663	193.691420	215.443551
69	146.056776	161.896937	179.789380	200.017934	222.906858
70	149.977911	166.539618	185.284114	206.518427	230.594064
71	153.977469	171.286759	190.916217	213.197684	238.511886
72	158.057019	176.140711	196.689122	220.060621	246.667242
73	162.218159	181.103877	202.606351	227.112288	255.067259
74	166.462522	186.178714	208.671509	234.357876	263.719277
75	170.791773	191.367735	214.888297	241.802717	272.630856
76	175.207608	196.673509	221.260504	249.452292	281.809781
77	179.711760	202.098663	227.792017	257.312230	291.264075
78	184.305996	207.645883	234.486818	265.388316	301.001997
79	188.992115	213.317916	241.348988	273.686495	311.032057
80	193.771958	219.117569	248.382713	282.212873	321.363019
81	198.647397	225.047714	255.592280	290.973727	332.003909
82	203.620345	231.111288	262.982087	299.975505	342.964026
83	208.692752	237.311292	270.556640	309.224831	354.252947
84	213.866607	243.650796	278.320556	318.728514	365.880536
85	219.143939	250.132939	286.278570	328.493548	377.856952
86	224.526818	256.760930	294.435534	338.527121	390.192660
87	230.017354	263.538051	302.796422	348.836617	402.898440
88	235.617701	270.467657	311.366333	359.429624	415.985393
89	241.330055	277.553179	320.150491	370.313938	429.464955
90	247.156656	284.798126	329.154253	381.497572	443.348904
91	253.099789	292.206083	338.383110	392.988755	457.649371
92	259.161785	299.780720	347.842687	404.795946	472.378852
93	265.345021	307.525786	357.538755	416.927834	487.550217
94	271.651921	315.445117	367.477223	429.393350	503.176724
95	278.084960	323.542632	377.664154	442.201667	519.272026
96	284.646659	331.822341	388.105758	455.362213	535.850186
97	291.339592	340.288344	398.808402	468.884673	552.925692
98	298.166384	348.944831	409.778612	482.779002	570.513463
99	305.129712	357.796090	421.023077	497.055424	588.628867
100	312.232306	366.846502	432.548654	511.724449	607.287733

AMOUNT OF ANNUITY $[(1 + i)^n - 1]/i$ (Continued)

Years n	Rate i				
	.035(3½ %)	.04(4 %)	.045(4½ %)	.05(5 %)	.055(5½ %)
1	1.00000000	1.00000000	1.00000000	1.00000000	1.00000000
2	2.03500000	2.04000000	2.04500000	2.05000000	2.05500000
3	3.10622500	3.12160000	3.13702500	3.15250000	3.16802500
4	4.21494288	4.24646400	4.27819113	4.31012500	4.34226638
5	5.36246588	5.41632256	5.47070973	5.52563125	5.58109103
6	6.55015218	6.63297546	6.71689166	6.80191281	6.88805103
7	7.77940751	7.89829448	8.01915179	8.14200845	8.26689384
8	9.05168677	9.21422626	9.38001362	9.54910888	9.72157300
9	10.3684958	10.5827953	10.8021142	11.0265643	11.2562595
10	11.7313932	12.0061071	12.2882094	12.5778925	12.8753538
11	13.1419919	13.4863514	13.8411788	14.2067872	14.5834982
12	14.6019616	15.0258055	15.4640318	15.9171265	16.3855907
13	16.1130303	16.6268377	17.1599133	17.7129828	18.2867981
14	17.6769864	18.2919112	18.9321094	19.5986320	20.2925720
15	19.2956809	20.0235876	20.7840543	21.5785636	22.4086635
16	20.9710297	21.8245311	22.7193367	23.6574918	24.6411400
17	22.7050157	23.6975124	24.7417069	25.8403664	26.9964027
18	24.4996913	25.6454129	26.8550837	28.1323847	29.4812048
19	26.3571805	27.6712294	29.0635625	30.5390039	32.1026711
20	28.2796818	29.7780786	31.3714228	33.0659541	34.8683180
21	30.2694707	31.9692017	33.7831368	35.7192518	37.7860755
22	32.3289022	34.2479698	36.3033780	38.5052144	40.8643097
23	34.4604137	36.6178886	38.9370300	41.4304751	44.1118467
24	36.6665282	39.0826041	41.6891963	44.5019989	47.5379983
25	38.9498567	41.6459083	44.5652101	47.7270988	51.1525882
26	41.3131017	44.3117446	47.5706446	51.1134538	54.9659805
27	43.7590602	47.0842144	50.7113236	54.6691264	58.9891094
28	46.2906273	49.9675830	53.9933332	58.4025828	63.2335105
29	48.9107993	52.9662863	57.4230332	62.3227119	67.7113535
30	51.6226773	56.0849378	61.0070697	66.4388475	72.4354780
31	54.4294710	59.3283353	64.7523878	70.7607899	77.4194293
32	57.3345025	62.7014687	68.6662452	75.2988294	82.6774979
33	60.3412101	66.2095274	72.7562263	80.0637708	88.2247603
34	63.4531524	69.8579085	77.0302565	85.0669594	94.0771221
35	66.6740127	73.6522249	81.4966180	90.3203074	100.251364
36	70.0076032	77.5983138	86.1639658	95.8363227	106.765189
37	73.4578693	81.7022464	91.0413443	101.628139	113.637274
38	77.0288947	85.9703363	96.1382048	107.709546	120.887324
39	80.7249060	90.4091497	101.464424	114.095023	128.536127
40	84.5502777	95.0255157	107.030323	120.799774	136.605614
41	88.5095375	99.8265363	112.846688	127.839763	145.118923
42	92.6073713	104.819598	118.924789	135.231751	154.100464
43	96.8486293	110.012382	125.276404	142.993339	163.575989
44	101.238331	115.412877	131.913842	151.143006	173.572669
45	105.781673	121.029392	138.849965	159.700156	184.119165
46	110.484031	126.870568	146.098214	168.685164	195.245719
47	115.350973	132.945390	153.672633	178.119422	206.984234
48	120.388257	139.263206	161.587902	188.025393	219.368367
49	125.601846	145.833734	169.859357	198.426663	232.433627
50	130.997910	152.667084	178.503028	209.347996	246.217476

INTEREST TABLES (Continued)

AMOUNT OF ANNUITY $[(1 + i)^n - 1]/i$ (Continued)

Years n	.06(6 %)	.065(6½ %)	.07(7 %)	.075(7½ %)	.08(8 %)
			Rate i		
1	1.00000000	1.00000000	1.00000000	1.00000000	1.00000000
2	2.06000000	2.06500000	2.07000000	2.07500000	2.08000000
3	3.18360000	3.19922500	3.21490000	3.23062500	3.24640000
4	4.37461600	4.40717463	4.43994300	4.47292188	4.50611200
5	5.63709296	5.69364098	5.75073901	5.80839102	5.86660096
6	6.97531854	7.06372764	7.15329074	7.24402034	7.33592904
7	8.39383765	8.52286994	8.65402109	8.78732187	8.92280336
8	9.89746791	10.0768565	10.2598026	10.4463710	10.6366276
9	11.4913160	11.7318522	11.9779887	12.2298488	12.4875578
10	13.1807949	13.4944225	13.8164480	14.1470875	14.4865625
11	14.9716426	15.3715600	15.7835993	16.2081191	16.6454875
12	16.8699412	17.3707114	17.8884513	18.4237280	18.9771265
13	18.8821377	19.4998076	20.1406429	20.8055076	21.4952966
14	21.0150659	21.7672951	22.5504879	23.3659207	24.2149203
15	23.2759699	24.1821693	25.1290220	26.1183647	27.1521139
16	25.6725281	26.7540103	27.8880536	29.0772421	30.3242830
17	28.2128798	29.4930210	30.8402173	32.2580352	33.7502257
18	30.9056525	32.4100674	33.9990325	35.6773879	37.4502437
19	33.7599917	35.5167218	37.3789648	39.3531919	41.4462632
20	36.7855912	38.8253087	40.9954923	43.3046813	45.7619643
21	39.9927267	42.3489537	44.8651768	47.5525324	50.4229214
22	43.3922903	46.1016357	49.0057392	52.1189724	55.4567552
23	46.9958277	50.0982420	53.4361409	57.0278953	60.8932956
24	50.8155774	54.3546278	58.1766708	62.3049874	66.7647592
25	54.8645120	58.8876786	63.2490377	67.9778615	73.1059400
26	59.1563827	63.7153777	68.6764704	74.0762011	79.9544151
27	63.7057657	68.8568772	74.4838233	80.6319162	87.3507684
28	68.5281116	74.3325743	80.6976909	87.6793099	95.3388298
29	73.6397983	80.1641916	87.3465293	95.2552582	103.965936
30	79.0581862	86.3748640	94.4607863	103.399403	113.283211
31	84.8016774	92.9892302	102.073041	112.154358	123.345868
32	90.8897780	100.033530	110.218154	121.565935	134.213537
33	97.3431647	107.535710	118.933425	131.683380	145.950620
34	104.183755	115.525531	128.258765	142.559633	158.626670
35	111.434780	124.034690	138.236878	154.251606	172.316804
36	119.120867	133.096945	148.913460	166.820476	187.102148
37	127.268119	142.748247	160.337402	180.332012	203.070320
38	135.904206	153.026883	172.561020	194.856913	220.315945
39	145.058458	163.973630	185.640292	210.471181	238.941221
40	154.761966	175.631916	199.635112	227.256520	259.056519
41	165.047684	188.047990	214.609570	245.300759	280.781040
42	175.950545	201.271110	230.632240	264.698315	304.243523
43	187.507577	215.353732	247.776496	285.550689	329.583005
44	199.758032	230.351725	266.120851	307.966991	356.949646
45	212.743514	246.324587	285.749311	332.064515	386.505617
46	226.508125	263.335685	306.751763	357.969354	418.426067
47	241.098612	281.452504	329.224386	385.817055	452.900152
48	256.564529	300.746917	353.270093	415.753334	490.132164
49	272.958401	321.295467	378.999000	447.934835	530.342737
50	290.335905	343.179672	406.528929	482.529947	573.770156

CHAPTER 8

Accrued Interest

You saw in Chapter 6, on bond price calculation, that the bond price equation includes a term for accrued interest and that, for municipal securities, the Municipal Securities Rulemaking Board (MSRB) prescribes a method for calculating accrued interest. That chapter did not consider the actual accrued interest calculation. This chapter considers the various methods of computing accrued interest, when they apply, and to what securities. When you finish this chapter, you should understand the concept of accrued interest, how to compute accrued interest, when it applies in a bond trade, and the different conventions in calculating accrued interest.

WHAT IS ACCRUED INTEREST?

Accruing interest on bond trades is simply a trade custom. A trade does not need to be done in this way. We could simply compute price using the first two terms in the equation given in Chapter 6, and in some countries this is done. In the United States, we add accrued interest to the price, and the buyer pays the total, consisting of two separate parts. For trades done at a dollar price, the accrued interest would be added, and the total becomes the total amount paid. However, the accrued interest could also be included in the bond price. This trade custom probably has a historical basis, possibly dating back for centuries.

Most investors aren't much concerned with calculating accrued interest. Computer programs do this, and the amounts aren't usually very large compared with the principal amount due. However, you should be aware of accrued interest. In the past, several scams have occurred because the participants in the trades didn't pay sufficient attention to the accrued interest part of the trades and the risks they were running. The largest and best known involved repurchase agreements with one of the country's largest and best known banks. It resulted in a considerable loss of both money and reputation to the bank.

BONDS THAT ACCRUE INTEREST

Not all bonds accrue interest. Bonds that are in default are traded at a dollar price, without accrued interest. This is because since the bonds are in default, they are not paying either principal or interest. How could anyone accrue interest when none is being paid? If you buy a bond that is traded flat, your salesperson should tell you about it. If you do this kind of transaction at all, you are probably already aware of the risks you are running. But any investor should certainly know whether or not the prospective bond investment is up to date in its interest payments; we say that it is current in its interest payments. If it isn't, the bond may well be trading flat.

THE EQUATION FOR ACCRUED INTEREST

The equation for accrued interest is the same as the equation for interest presented in Chapter 2:

$$I = Sit \qquad\qquad \text{Equation 8.1}$$

In this case,

I = accrued interest
S = principal amount, the par amount of the bonds in this case
i = coupon rate of the bonds
t = the time during which the bonds have accrued interest since the last interest payment or since the dated date of the bond

The dated date is almost always the date the bond starts paying interest.

I, S, and i are almost always clear from the problem. The time is set by the rules governing the particular trade under consideration. Two different rules exist in the United States. You should know about both rules.

The settlement period is the period between interest payments in which the trade actually settles. This is usually, but not always, the same period in which the trade takes place.

Suppose you have a bond that makes interest payments each February 1 and August 1. You make a trade of this bond on September 19 for settlement on September 23. The settlement period is August 1 to February 1, and both trade and settlement dates are in the same period. But suppose you trade on July 29 for settlement August 2. The trade date and the settlement date are in two different settlement periods.

RULE 1: 30-DAY MONTH, 360-DAY YEAR

In this rule, each month is assumed to have 30 days, and so a year will have 360 days. Thus, the periods from February 1 to March 1, from March 1 to April 1, and from April 1 to May 1 each have 30 days.

The equation for the day count is as follows:

$$\text{Number of days} = (Y2 - Y1)360 + (M2 - M1)30 + (D2 - D1)$$

In this calculation, $Y1$, $M1$, and $D1$ are the year, month, and day, respectively, that the computation period begins, and $Y2$, $M2$, and $D2$ and the year, month, and day, respectively, that the computation period ends. (Source: Municipal Securities Rulemaking Board Rule Book.)

Suppose the $D1$ or $D2$ is 30 or 31? If $D2$ has a value 31, and $D1$ has a value 30 or 31, then $D2$ is changed to value 30. If $D1$ has a value 31, then $D1$ is changed to value 30. In addition, the time periods specified by this rule (Rule G-33), include the day specified in the rule at the beginning of the period, but not the day specified at the end of the period. (Source: Municipal Securities Rulemaking Board Rule Book, slightly modified.)

Corporate, municipal, and federal agency bonds use this rule, as do some municipal notes.

RULE 2: EXACT DAYS OVER EXACT DAYS

In this case, the exact number of days since the last interest payment period is divided by the exact number of days in the settlement period.

United States Treasury bonds and notes and some municipal notes, use this rule. If you are trading some municipal notes, how will you know which rule is followed? The description of the notes or the municipal note trader will tell you.

In this case, the numerator for calculating t is the total number of days since the last interest payment, or the day the security started paying interest, which is almost always the dated date. This day count includes Saturdays, Sundays, and holidays. The denominator is the total number of days in the interest payment period. This could vary for different securities on the same day, and for any given security at different times in the year.

CHAPTER SUMMARY

The equation for accrued interest is as follows:

$$I = Sit$$

where I = accrued interest
  S = par amount
  i = coupon rate of the bonds
  t = the time during which the bonds have accrued interest since the last interest payment or since the dated date of the bond

The two main conventions for computing accrued interest are (1) 30-day month, 360 day year (used for municipals, corporates, federal agencies, and some municipal notes) and (2) exact days over exact days (used for Treasuries and some municipal notes).

COMPUTER PROJECT

Compute the accrued interest, using each convention, for different combinations of accrued interest start date and settlement date. Are there any combinations that result in the same accrued interest? What is the largest possible difference in results between the two conventions?

TOPIC FOR CLASS DISCUSSION

Discuss the two main day count conventions. Which do you prefer and why? Can you design another convention that might better solve the day count problem?

PROBLEMS

1. You have just bought the $10,000 par amount of a 6% municipal bond for delivery on December 15, with interest payable February 15 and August 15. What is your accrued interest?

2. You have just bought the $10,000 par amount of a 6% Treasury bond for delivery on December 15, with interest payable February 15 and August 15. What is your accrued interest?

3. You have just bought another $10,000 of each of the securities in Problems 1 and 2 for delivery on June 15. How much is your accrued interest in each case? What question must you answer before you can solve the problem?

CHAPTER 9

Discount Yield

The previous chapters looked at the equations for compound interest and its related topics: present value, annuity certain, bond price, and the future value of an annuity. This chapter looks at discount yield and how to compute prices using discount yield. It shows the equation for price, given discount yield and some examples of securities traded using discount yield, especially United States Treasury bills. It develops the equation for bond equivalent yield, and shows why bond equivalent yield is important and where it is used.

United States Treasury bills (T-bills) are arguably the most important short-term securities issued anywhere in the world, and possibly the most important securities of any sort issued anywhere. You should know about T-bills, how they are quoted, issued, and traded. You should understand how to compare T-bill discount yields with the bond yield that we studied in the first chapters of this book. We use T-bills as the example for discount yield calculations.

When you finish this chapter, you should understand the definition of discount yield, how to compute price from discount yield, the relation between discount yield and bond equivalent yield, when and why you should use bond equivalent yield, and the most important securities (T-bills) traded on a discount yield basis.

DISCOUNT YIELD

Discount yield is used in many cases instead of bond yield to compute security prices. Almost always, discount yield is used to compute prices of relatively short-term securities, usually securities maturing in 1 year or less. Years ago, lenders and financial books used the phrase "the interest is deducted in advance."

Discount yield is expressed as an annual rate. The discount yield is used to compute the discount (amount), which in turn is used to compute the price of the security.

Discount yield is not the same as bond yield, or the interest rate i, which we discussed in the first seven chapters of this book. We will discuss the relationship between discount yield and bond yield later in this chapter, in the section on bond equivalent yield.

CALCULATION OF THE DISCOUNT AND PRICE
FOR DISCOUNT SECURITIES

Suppose we denote the discount by D and the discount rate (or yield), expressed as a percent and time by d and t. Then

$$D = d \times t \qquad \text{Equation 9.1}$$

Compute the price by subtracting the discount from 100, so

$$P = 100 - D \qquad \text{Equation 9.2}$$

P is the price of the security.

WHAT IS A TREASURY BILL?

A Treasury bill, or T-bill, is a short-term discount security issued by the United States Treasury. T-bills are sold at a discount and do not bear a stated rate of interest. Instead, interest is earned when the bills mature at par or are sold at a profit. This profit is considered interest income and is reported and taxed as interest income on tax returns.

The most common T-bills are the 3-month and 6-month T-bills. These are auctioned every Monday and delivered and paid for on the following Thursday. The Federal Reserve acts as an agent for the U.S. Treasury in this auction. The 3-month bills mature 91 days after the Thursday they are delivered and paid for, and the 6-month bills mature 182 days after the Thursday for delivery and payment. The Treasury used to issue 1-year bills but discontinued offering these. Occasionally, the Treasury issues cash management bills. Cash management bills are short-term bills designed to help the Treasury manage its cash flow.

As an example, on Monday, September 24, 2001, the Treasury sold an issue of 3-month bills and an issue of 6-month bills for delivery on Thursday, September 27, 2001. The discount yields and prices were 2.380% and 99.398 for the 3-month bills, and 2.360% and 98.807 for the 6-month bills. The bills matured on December 27, 2001, and March 28, 2002, respectively. A buyer of $10,000 par amount of the 3-month bills would pay $9,939.80 for these bills. If held until maturity, the buyer would receive $10,000 and have interest earnings of $60.20.

THE DAY-COUNT CONVENTIONS FOR T-BILLS

T-bill discounts (and prices) are calculated using a year of 360 days (the denominator) and counting all days from purchase date to sale or maturity date, including Saturdays, Sundays, and holidays in the day count calculation (the numerator). For example, the period from February 1 to February 15 is 14 days. The period from February 1 to March 1 is 28 days, except in leap years, when it is 29 days. The period from March 3 to May 17 is 75 days, as follows:

March 3 to 31 = 31 − 3 =	28
April has 30 days	30
May has 17 days to May 17	17
Total:	75

For T-bill calculations, a year has 360 days, so the denominator in the time calculation is 360. However, in computing the days to maturity, all days are counted. If you buy a security on January 15 that matures on January 15 next year, the numerator in the calculation will have the actual day count, 365 days, except in leap years, when it will have a count of 366 days.

EXAMPLE 6.1.   In the case of the 3-month bills discussed here, compute the day count and price as follows:

September 27 to 30 = 30 − 27 =	3
October has 31 days	31
November has 30 days	30
December 1 to 27 =	27
Total:	91

The time $t = 91/360$
The discount rate $d = 2.380\%$
The discount $D = (d) \times (t) = (2.380)(91/360)$
$$= (2.380)(.252778)$$
$$= .602$$
Price $P = 100 - D = 100 - .602$
$$= 99.398, \text{ as reported and shown earlier} \quad \blacksquare$$

EXAMPLE 6.2.   Compute the price for the 6-month bills as follows:

$t\ = 182/360$
$d\ = 2.360$
$D = (2.360)(182/360) = 1.193$
$P\ = 100 - 1.193$
$\quad = 98.807, \text{ as reported and shown earlier} \quad \blacksquare$

PRICE CALCULATIONS FOR DISCOUNT MUNICIPAL SECURITIES

For discount municipal securities, the Municipal Securities Rulemaking Board has its own rule, shown in Rule G-33. It is similar to the equation presented earlier for T-bills, but it uses the municipal securities day count convention of a 30-day month and a 360-day year. This is discussed more fully in Chapter 8 on accrued interest.

BOND EQUIVALENT YIELD (BEY): WHAT IT MEANS AND HOW TO COMPUTE IT

Suppose you buy a 1-year discount security at a 10% discount yield and hold it to maturity. How much will you have earned, as a true annual yield and as a nominal annual rate, compounded semiannually?

You will have earned a true annual rate of 11.11%, as follows:

Investment: 90
Gain (interest income) $= \underline{100}$ (maturity amount) $- 90$ (cost)
$\qquad\qquad\qquad = \ \ 10$

You earned this 10 on an investment of 90, so the percentage return = 10/90 = 11.11%.

Expressing this as a nominal annual rate, compounded semiannually, we have 10.8184%, or 5.4092% per half-year.

We can check this:

$$(90) \times (1.054092) \times (1.054092) = 100$$

The interest rate 10.8184% is called the "bond equivalent yield" of an annual discount rate of 10.00. It is the rate that is equivalent in interest earnings to the discount rate. It is mathematically equivalent to the Y in the bond price calculations shown in Chapter 6.

DERIVATION OF THE BOND EQUIVALENT YIELD EQUATIONS

Let d be the annual discount rate and y the bond equivalent yield, both expressed as decimals. We have the purchase price, compounded back to a maturity value of 1, and we want to solve for y in terms of d:

For one year,

$$(1-d)\left(1+\frac{y}{2}\right)^2 = 1$$

Then

$$\left(1+\frac{y}{2}\right)^2 = \frac{1}{1-d}$$

$$\left(1+\frac{y}{2}\right) = \frac{1}{(1-d)^{1/2}}$$

$$y = 2\left(\frac{1}{(1-d)^{1/2}} - 1\right)$$

We can generalize to time t, t greater than one-half year and expressed as a decimal:

$$(1-td)\left(1+\frac{y}{2}\right)^{2t} = 1$$

$$\left(1+\frac{y}{2}\right)^{2t} = \frac{1}{(1-td)}$$

$$\left(1+\frac{y}{2}\right) = \frac{1}{(1-td)^{1/2t}}$$

$$y = 2\left[\left(\frac{1}{(1-td)^{1/2t}}\right) - 1\right]$$

For securities maturing in 6 months or less, we use the equation for simple interest, because we have no compounding. We have (d, t, and y expressed as decimals)

$$(1 - dt)(1 + yt) = 1$$ Equation 9.3

Multiplying out and simplifying, we have

$$1 - dt + yt - dyt^2 = 1$$
$$yt - dyt^2 = dt$$
$$y - dyt = d$$

and we obtain

$$y = \frac{d}{1 - dt}$$ Equation 9.4

Suppose we have T-bills, with their convention of a 360-day year. In that case, we use the 360-day year convention and have

$$t = \frac{t_d}{360}$$

expressed as a decimal, where t_d is the number of days until maturity. We then have

$$y = \frac{d}{1 - d\dfrac{t_d}{360}} \bullet \frac{365}{360}$$ Equation 9.5

where the last factor 365/360 converts the yield to a 365-day year.

WHY WE CARE ABOUT BOND EQUIVALENT YIELD (BEY)

We use BEY to compare the yields of discount securities and securities traded on a bond yield basis to determine which offers the higher yield.

EXAMPLE 6.3.   In the previous case, with the 1-year 10% discount security, suppose you are offered a 1-year T-bill at a 10% yield and a 1-year Treasury note at a 10.40% yield. Your broker urges you to buy the 1-year note because "it offers you a better yield." You calculate the BEY for the bill, just as we did earlier, and discover that the BET is 10.81%. You decide to buy the bill because it offers over 40 basis points higher yield than the 1-year note.

∎

EXAMPLE 6.4. (from Chapter 1) Your broker has offered you a 6-month T-bill (maturing in 182 days) at a 3.92 discount yield and a Treasury note maturing in 6 months at a bond yield of 4.00. You compute the BEY as follows, using equation 9.5:

$$BEY = [.0392/(1 - \{.0392\}\{182/360\})][365/360]$$
$$= [.0392/(1 - \{.0392\}\{.5055556\})][365/360]$$
$$= [.0392/(1 - .01988178)][1.013889]$$
$$= [.0392/.98011822][1.013889]$$
$$= [.03999518][1.013889]$$
$$= 4.055$$

You buy the T-bill because it yields 5.5 basis points more than the note.

■

A HISTORICAL NOTE

Most leading papers report T-bill yields. Years ago, these reports showed only the discount bid and ask yields. Then, some time ago, the market regulators required these reports to show bond equivalent yield as well, and most market reports now also show these yields. They show them in a separate column, variously titled "Bond equivalent yield," "Bond yield," "Investment rate," or simply "Yield." The regulators wanted prospective investors in T-bills easily to be able to find out the true yield of the bills they might have an interest in buying. This example shows a case where the securities regulators caused market reporters to provide more, and important, information to investors.

TAXATION OF INCOME FROM TREASURY BILLS

Generally, gains from investment in T-bills, whether from maturing bills or from sale before maturity, may be reported and taxed as interest income for federal income tax purposes. Always consult with your tax adviser in these matters.

CHAPTER SUMMARY

Discount yield is used to price certain securities, usually short-term securities.
Treasury bills (T-bills) are the most important of these securities.

The equations for discount and price are

$$D = d \times t$$
$$P = 100 - D$$

Where d = discount yield (or rate)
t = time to maturity
D = discount
P = price of the security

Bond equivalent yield (BEY) is the bond yield equivalent of the discount yield.

Bond equivalent yield is used to compare securities trading at a discount yield with securities trading at a bond yield.

The equations for BEY are as follows. For securities maturing in more than 6 months:

$$y = 2\left[\left(\frac{1}{(1-td)^{1/2t}}\right) - 1\right]$$

For securities maturing in 6 months or less:

$$y = \frac{d}{1 - d\frac{t_d}{360}} \bullet \frac{365}{360},$$

where d = discount yield
y = bond yield
t = time to maturity
t_d = time to maturity in days

COMPUTER PROJECTS

1. Use your favorite mathematical computer program to compute or display the bond equivalent yield for a noninterest-bearing security, like a T-bill, maturing in 1 year, at a 10% discount yield, and the difference between the BEY and the discount yield. Now vary the discount yields and show the BEYs for a variety of discount yields, and the differences between the discount yields and their respective BEYs. Does the difference increase or decrease as you lower the discount yield? Why? Describe why this should happen.
2. Now vary the time to maturity for a particular discount yield and its equivalent BEY. What happens to the difference as the time to

maturity decreases and as the time to maturity increases? Why? Describe why this should happen.

TOPICS FOR CLASS DISCUSSION

1. What advantages and disadvantages did the use of discount yield have before the widespread use of computers?
2. What advantages and disadvantages does the present use of discount yield have? Would you recommend changing this trade custom? What would you recommend as a replacement and why?
3. In the case of the 90-day bills sold on September 24, 2001, the market report showed a BEY, called "investment rate," of 2.429%. How many decimal places was this accurate to? Would you consider this enough accuracy?

PROBLEMS

1. On Monday, April 22, 2002, the Treasury sold some 3-month bills at a yield of 1.690%. What was the price and the bond equivalent yield for these bills?
2. On the same day, the Treasury also sold some 6-month bills at a yield of 1.880%. What was the price and the bond equivalent yield for these bills?
3. Your bond salesperson has offered you a six-month Treasury note yielding 1.90% and says that you should buy the note because it offers a higher yield. What is your reply?

CHAPTER 10

Calculations for Other Securities

Many other types of fixed-income securities exist. Almost all of them use the equations and methods we have discussed earlier in this book, either the present value equation of the sort used to price bonds, or a discount yield of the sort used to price Treasury bills. When you examine the security, you should be able to figure out the mathematical equation for pricing it.

Here are two examples of other types of fixed-income securities, using pricing methods discussed earlier.

When you finish this chapter, you should have some understanding of how other types of securities might be priced, with these two examples.

CERTIFICATES OF DEPOSIT

A certificate of deposit (CD) represents a bank deposit. It has an interest rate, an amount, and a maturity date, so you can apply the standard calculation methods for bonds. A few CDs do not have a stated interest rate but are sold at a discount. To price these CDs, use the equations for T-bills.

CDs, of the sort we are considering here, are issued in multiples of $100,000. Because of their size, Federal Deposit Insurance Corporation Insurance does not apply, except to the first $100,000. They have a secondary market, and the owner can sell them if desired.

Years ago, all banks were considered to have about the same credit rating, but these days banks may have different credit ratings. As a result, CDs issued by different banks may trade at different prices. This practice is called "tiering."

CDs pay interest at maturity if the original term is less than 1 year, and they pay interest semiannually if the original term is 1 year or greater. Thus, for interest-bearing CDs, use the bond price equations to compute price from yield and add the accrued interest. For discount CDs, use the equation for a T-bill to compute the price.

REPURCHASE AGREEMENTS

A repurchase agreement (repo) is the sale of a security combined with an agreement to buy back (repurchase) the security at a specified time and price. The price is set to give the temporary owner of the security a satisfactory yield.

Economically, a repo is a loan, but legally, it is a sale. The buyer is the legal owner of the security. If the repurchaser cannot buy back the security, the buyer remains the actual legal owner. Therefore, the credit of the repurchaser is important to the temporary buyer, and buyers must also make sure that the market value of the securities they are buying is greater than the amount they are paying for them. Most large buyers in the repo market have credit departments that investigate the credit of the original seller and must approve this credit before any repurchases are done. In the past, occasional bankruptcies have forced original buyers actually to sell the security they had bought.

A reverse repo is a repo from the point of view of the original buyer.

Repos may, and have, been done with many different fixed-income securities. These include United States Treasuries, federal agencies, mortgage-backed pass-throughs, municipals, and some money-market instruments. Repos can be put together by repo brokers or by the buyer and seller doing business directly. They may go for 1 day to 1 year or more, and some are simply left open.

Many lenders cannot lend unsecured, either by law or by the agreement that sets them up. But repos are secured. The only question is whether the market value of the repo is greater than the amount actually paid for it. This requires that temporary buyers make sure that they continue to monitor the market prices of the securities they own on repos.

USES OF REPOS AND REVERSE REPOS

Security dealers use repos to finance their security inventories. Sometimes large investors, such as pension funds, may use repos to earn extra income. These investors sometimes can lend securities directly for a fee, beyond the usual interest income or dividends paid by the security. Repos can also directly provide short-term financing for the business.

Reverse repos provide financing to other firms and give the reverse repos provider a safe and profitable investment and earn income. They can also be used to cover short sales. The securities must be returned, but meanwhile the short sale is covered.

PRICING REPOS

Repo sale and repurchase prices are set to give the temporary owner his or her required yield.

CHAPTER SUMMARY

Many fixed-income securities exist. These may be priced using bond yield or discount yield. Use either the bond price calculation equations, discussed in Chapter 6, or the discount price calculation equation, discussed in Chapter 9, to price these securities. The choice depends on the particular pricing custom for the security concerned.

CLASS PROJECT

Look at a variety of fixed-income securities, such as bankers acceptances, federal funds, and others. How are these priced? Would you suggest any changes in the methods of pricing these securities?

CHAPTER 11

Quotations and Bond Market Reports

If you do any work at all in finance, you should know how to read a bond market report. Yet every class I have ever given has had students who had not learned this skill. Although it isn't really mathematical, it is an important part of the skills needed to work effectively in any financial field, so it deserves a place in this book. When you finish this chapter, you should be able to read just about any fixed-income market report with understanding.

Remember that bond prices are stated as a percent of par, not in dollars. However, market quotations are frequently stated as a yield basis. The yield must be converted to a dollar price (percentage of par) in order for the actual transaction to be done. Use the equations shown in Chapter 6 or Chapter 9 to do this.

LONG-TERM INSTRUMENTS

Long-term instruments, including most Treasury bonds and notes, federal agency bonds, corporate bonds, and municipal term bonds, are generally quoted on a dollar price basis, as a percent of par. Treasuries are quoted as a percent of par, frequently with the 32s after a decimal point. The minimum change for Treasuries and federal agencies is generally $\frac{1}{32}$ of a point. Municipal term bonds and corporate bonds generally have a minimum change of $\frac{1}{8}$ point. A municipal term bond is, generally, a long-term bond with a sinking fund, which may be active, and a call feature.

Almost all bonds are traded in a dealer market. An active market exists in Treasury bonds, with active bids and offerings in most Treasury securities.

EXAMPLE 11.1. In late April of 2002, the market report for the longest term (nonindexed) Treasury bond might read as follows:

Feb   31   k   5   3/8   96.21   96.22   +0.24   5.61

This means, entry by entry:

Feb 31	The bond matures in February 2031.
k	This is a Treasury bond; nonresident aliens are exempt from withholding tax.
5 3/8	The bond has a coupon rate of 5 3/8 (5.375)%.
96.21	The bond has a reported bid of 96 and 21/32s percent of par. This equals 96.65625% of par.
96.22	The bond has a reported ask price of 96 and 22/32s percent of par. This equals 96.6875% of par.
+0.24	The closing price, at the time this report was compiled, was 24/32s (.75) higher than the price the previous day
5.61	The bond's yield to maturity was 5.61%, nominal annual rate, compounded semiannually, as usual in the United States.

Note that the 96.21 and 96.22 quotes are a dealer's bid and ask for this bond. The dealer will pay 96.21 for the bond and offer it at 96.22. The dealer's operating profit is the spread between the bid and the ask price. Corporate bonds and dollar-quoted municipal bonds are reported in a similar manner. ∎

EXAMPLE 11.2 In late April of 2002, a corporate bond report in Stock Exchange Bond trading might read as follows:

ATT   8 5/8s   31   8.8   213 $98\frac{1}{2}$ $-1\frac{1}{2}$

This means, entry by entry:

ATT	The issuer is the AT&T Corporation.
8 5/8s	The coupon rate is 8 5/8%.
31	The bonds mature in 2031.
8.8	The current yield (8.625/98.5 = 8.8).
$98\frac{1}{2}$	The closing price.
$-1\frac{1}{2}$	The change from the previous close.

Another report, at about the same time, might show the following:

Virginia El&Pwr 5.75's 06 A3/A- 100.68 100.86 5.53 −0.09

This means, term by term:

Virginia Electric & Power is the issuer.
5.75   The coupon rate of the bonds.
06   The bonds mature in 2006.
A3/A-   The bonds have a Moody's rating of A3 and a S&P rating of A-.
100.68   The bid price is 100.68.
100.86   The ask price is 100.86.
5.53   The yield to maturity is 5.53%, nominal annual rate compounded semiannually.
−0.09   The closing price was down .09 from the previous close.

Some long-term municipal bonds, when-issued Treasuries, and a very few others are quoted in a yield basis. When-issued Treasuries are Treasuries scheduled to be auctioned in a few days. The auction has not been held yet, so no one knows the coupon rate, and a yield basis is the only way to trade these when-issued securities. After the auction, the results are known, the new securities will receive a coupon rate, and future trading will take place on a dollar price basis.

When-issued securities are securities that will be offered, or have been sold but not yet delivered to the buyers. ∎

EXAMPLE 11.3.   The following might be a municipal bond quotation in late February of 2002:

Waterloo CSD, NY 4.50s 06/15/20 @ 4.80 Ca 06/15/11@100

This means, item by item:

Waterloo CSD, NY   The issuer is Waterloo Central School District, New York.
4.50   The coupon rate is 4.50%.
06/15/20   The bonds mature on June 15, 2020.
4.80   The bonds are offered at a yield of 4.80 to maturity.
Ca 06/15/11@100   The bonds may be called on June 15, 2011, at par, and after, according to the terms of the bond agreement. ∎

DISCOUNT INSTRUMENTS

Discount instruments are usually quoted on a yield basis, but this can be either the discount yield, studied in Chapter 9, or the bond yield, studied in Chapter 6.

Treasury bills are quoted in discount yield, as discussed in Chapter 9. A market report in late April of 2002, for T-bills might read as follows:

Oct 24 02 1.85 1.83 −0.03 1.87

This means, item by item:

Oct 24 02	The bills mature on Oct 24, 2002.
1.85	The bid is 1.85 discount yield.
1.83	The offering is 1.83 discount yield.
−0.03	The yield is 0.03 basis points lower than the previous close.
1.87	The bond equivalent yield is 1.87%.

Treasury STRIPS and municipal zero coupon bonds are quoted in bond yield. For STRIPS, the actual bond price is also usually given at the same time.

STRIPS (Separately Traded Interest and Principal Securities) are Treasury obligations, either notes or bonds, that have been split into their separate interest and principal payments and sold as zero coupon instruments. The Treasury does not create STRIPS. Instead, it makes certain Treasuries eligible for the STRIPS program. Individual dealers then buy these eligible securities and break the up into STRIPS. STRIPS can also be put back together again to recreate the original securities, and dealers sometimes do this as well.

For example, a market report for an issue of STRIPS in late April of 2002, might read as follows:

Nov 11   ci   59.19   59.21   +14   5.48

These entries mean the following:

Nov 11	The STRIPS mature in November of 2011.
ci	The STRIPS are stripped coupon payments.
	Other codes are bp (bond principal) and np (note principal).
59.19	The bid for the STRIPS.
59.20	The ask price for the STRIPS.
+14	The change from the previous close, in 32s.
5.48	The yield to maturity for the STRIPS.

CHAPTER SUMMARY

Many long-term instruments, including Treasuries, agencies, corporates, and municipal term bonds, are quoted in dollar prices.

When-issued Treasuries and many long-term municipal bonds are quoted in bond yield.

Discount securities are quoted in yield. T-bills are quoted in discount yield. Treasury STRIPS and municipal zeros are quoted in bond yield.

CHAPTER 12

Types of Yields

Many times, the same word can have several different meanings depending on the context in which it is used. In finance, the world "yield" can have several meanings depending on its use or on the adjective used to modify it. Generally, yield means some kind of return on an investment of some kind. In its most general usage, it is not specific on either the return or the investment. For example, we may speak of farm yield as bushels of corn per acre, or bushels of wheat per acre, or simply that the farm yields "a good living." In semiconductors, we speak of the yield as the percentage of total output that meets required specifications.

In finance, the word "yield" may have at least four meanings depending on its use. They are nominal yield, current yield, discount yield, and true (or bond) yield. This chapter examines each of these meanings. You should understand these terms, what they mean, when and how to use them, how they relate to one another, and how to compute them. We have already discussed true, or bond, yield and discount yield.

When you finish this chapter, you should understand the meanings of these terms, how to compute them, and when they are used.

NOMINAL (OR COUPON) YIELD

The nominal yield is the coupon rate of the bond as a percent of par. For example, if a bond has a coupon of 6%, the nominal yield is 6%. This stays the same during the life of the bond, except for variable rate bonds.

DISCOUNT YIELD

Discount yield is used to compute the price of certain discount obligations, mostly United States Treasury bills. Although traded on a discount yield basis, many investors and portfolio managers convert the discount yield to the equivalent bond yield for management purposes. We discussed discount yield and bond equivalent yield in Chapter 9.

CURRENT YIELD

The current yield is the coupon rate as a percent of the price of the bond.

$$\text{Current yield} = (\text{Coupon rate})/(\text{Price})$$

or

$$CY = C/P \qquad\qquad \text{Equation 12.1}$$

EXAMPLE 12.1.   You buy a bond at 80, which has a coupon rate of 4%. Your current yield = 4/80 = 5%.

Current yield is easy to understand, use, and compute. You can apply current yield to an individual bond or to a portfolio of bonds. You can use current yield to make income comparisons. Under SEC regulations, current yield is used to show the return of bond mutual funds.

But current yield has several disadvantages, and you should understand these as well. It can show very misleading results if not used properly. For example, you buy a 12% bond due in 1 year at a price of 110. The current yield = 12/110 = 10.91%. Actually, your true return is around 2%. Current yield can give an appearance of accuracy that may not be actually present. For example, you can compute the current yield for a bond that has a high chance of being called or that is subject to a sinking fund call. This can lead to wrong investment decisions if you don't understand current yield and don't fully analyze the proposed investment.   ■

TRUE (OR BOND) YIELD

Bond yield is the yield, and, divided by 2, the interest rate i, and the internal rate of return, which we discussed in the first chapters of this book and which we used in computing bond prices in Chapter 6. It is used in most financial analysis and is the yield mostly used in this book. It is the standard for most financial analysis.

However, bond yield is hard to compute, given the price, coupon, and time to maturity. Some regulations require the calculation of yield, given the price. For example, regulations of the Municipal Securities Rulemaking Board require that, in the case of municipal bond trades done at a dollar price, the yield to maturity be shown, correct to .05%.

Bond yield is computed, given coupon, yield, and time to maturity, by using the standard price calculation equation. In most cases, this requires a numerical analysis technique. The bisection method, or some other technique, can be used for this calculation. This will require a decision on precision requirements. In some cases, such as in the municipal bond case mentioned previously, laws or regulations specify the precision requirements. In other cases, the analyst or the analyst's boss may specify the requirements.

True yield calculations use the present value calculations covered in the first six chapters of this book. Note that these equations assume reinvestment at the purchase yield rate of earnings not paid out in interest payments.

Using true yield has substantial advantages:

1. It is the standard formula, used in almost all analytical work in finance, insurance, and project analysis. In some cases, it is or may be required by law, regulation, or company policy.
2. It provides an accurate representation, within its assumptions, of the flow of earnings and return of capital investments.
3. It provides comparability between various investments or between various projects to help in making investment or management choices.
4. It has many analytical tools available, including a wide variety of textbooks, other books and periodicals, and computer programs.
5. It is widely used in a variety of fields besides finance, including actuarial work, engineering economy, and project analysis.

However, using true yield does have some problems:

1. It is a somewhat complicated equation, requiring the use of computers or tables.
2. One cannot go easily from price to yield.

True yield is the standard in almost all financial and other analysis. You should be thoroughly familiar with it and its uses.

A METHOD TO COMPUTE APPROXIMATE
BOND YIELD, GIVEN PRICE

To compute an approximate bond yield from price, divide average income by average investment. Average investment is the price plus par, divided by 2. Average income is the coupon rate, plus the discount divided by time to maturity, or minus the premium divided by time to maturity.

For premium bonds, we have the following:

$$\text{Average coupon} = C - \frac{Prem}{T} \qquad \text{Equation 12.2}$$

For discount bonds, we have

$$\text{Average coupon} = C + \frac{Disc}{T} \qquad \text{Equation 12.3}$$

Then average income will be

$$\text{Average income} = C \mp \frac{\left(\dfrac{PREM}{T}\right)}{\left(\dfrac{DISC}{T}\right)} \qquad \text{Equation 12.4}$$

$$\text{Average investment} = \frac{(Par + Price)}{2} \qquad \text{Equation 12.5}$$

Then for the estimate calculation, we have

$$\text{Yield} = \frac{\left(C \mp \left[\dfrac{\dfrac{PREM}{T}}{\dfrac{DISC}{T}}\right]\right)}{\dfrac{(Par + Price)}{2}} \qquad \text{Equation 12.6}$$

If the bond has a call feature, substitute

(Premium − Call premium) for Prem in the case of premium bonds
(Disc + Call premium) for Disc in the case of discount bonds, and
(Call price + Price) for (Par + Price)

This equation assumes that the owner earns equal annual income on an equal annual investment value.

The equation has the advantage that you can go from price to yield easily. You might want to do this as part of an approximation procedure, such as the bisection method, which requires a reasonable starting point. Or you might simply want to do a quick calculation with reasonable accuracy.

However, using this method, the yield differs from true yield increasingly as price differs from par. This error could eventually become large. In addition, this is not an industry standard. However, it can produce reasonably good estimations.

EXAMPLE 12.2.   The price of a 2% bond due in 10 years at a 6% yield = 70.245. Computing the yield back from the price, using the estimation, we have:

$$\text{Numerator} = 2 + (100 - 70.245)/10 = 4.98$$
$$\text{Denominator} = (100 + 70.245)/2 = 85.12$$
$$\text{Computed Yield} = 4.98/85.12 = 5.85$$

The computed yield differs from the true yield by 15 basis points. Depending on the particular situation, this could be accurate enough. It could also be used as a starting point to determine the yield with greater accuracy in some numerical analysis solutions, such as the bisection method. ■

CHAPTER SUMMARY

The world "yield" may have several meanings.

Nominal yield is the coupon rate of the bond.

Current yield is used in some cases. The equation for current yield (CY) is CY = C/P.

Discount yield is used to price some discount securities.

Bond yield is the yield used to compute bond prices and is standard for most financial analytical work.

It is difficult to go from price to yield.

Use numerical analysis techniques to compute yield from price.

An equation for the approximate bond yield from price is as follows:

$$\text{Yield} = \frac{\left(C \mp \left[\dfrac{\dfrac{PREM}{T}}{\dfrac{DISC}{T}} \right] \right)}{\dfrac{(Par + Price)}{2}}$$

COMPUTER PROJECT

Use your favorite mathematical program to compute yields given prices. Using Equation 12.6, show the calculated yield, true yield, and difference. Start with a coupon, yield, and time to maturity combination, such as in Example 12.2. Vary the time to maturity. What happens to the difference? Then vary the coupon rate and then the true yield. How do the differences between true yield and calculated yield change?

TOPICS FOR CLASS DISCUSSION

1. Under what conditions would you use Equation 12.6 in actual work?
2. Managers of bond mutual funds frequently use current yield to measure and advertise their performance. What results might this have on their portfolio management techniques? What would you look for if you were thinking of investing in bond mutual funds?

PROBLEMS

1. You have just bought a 2% bond, maturing in a little over 10 years, at a 10% yield and a price of 50. What are (a) the nominal yield, (b) the current yield, and (c) the true yield.
2. You own a 2% bond maturing in 3 years, a 5% bond maturing in 6 years, and a 9% bond maturing in 10 years. You swap your 2% bond for another 2% bond maturing in 20 years at a price of 90 for both bonds and equal maturity amounts. You swap your 5% bond for another 5% bond maturing in 25 years at a price of 95 for both bonds and equal maturity amounts, and you swap your 9% bond for another 9% bond maturing in 30 years at a price of 100 for both bonds and equal maturity amounts. What is the change in the current yield and the nominal yield of your portfolio and why?

CHAPTER 13

Sources of Return, Total Return, and Interest on Interest

This chapter discusses sources of return, how they are determined, the assumptions required to use the process, and the results one might expect from using the process. The concept of total return is defined and developed.

When you finish this chapter, you should understand the concept of total return, the assumptions you must make to compute it, how to compute it, and the situations where you might apply it.

Suppose you put aside some money into a fund and don't plan to take anything out of the fund for some time. You plan to reinvest any income the fund receives and liquidate the fund and take the proceeds at some future time. You may or may not add additional money to the fund before you liquidate it.

You can see that this situation resembles the examples we studied in earlier chapters on present values and the future value of an annuity. In the case of the future value of an annuity, we added fixed amounts to the fund, at fixed periodic intervals, and earned a fixed rate on the assets of the fund. In the case of present value, we set aside a fixed amount and earned at a fixed rate for a fixed period of time. In this chapter we generalize these concepts somewhat. Later in the book, we will generalize these further.

EXAMPLES

Here are some examples of such funds:

1. You set up a college fund for your new baby. You plan to deposit various amounts, at various times, into the fund. You will earn at varying rates over the life of the fund, which you expect to be 18 years.
2. You set up a retirement fund. You plan to deposit various amounts, at various times, with some amounts matched by your employer. You will earn varying rates over the life of the fund, which you expect to be 40 years.

We will look at one particular type of such a fund, where the investor makes an initial contribution, reinvests all interest earned, sets a date to close out the fund, and closes out the fund on the date planned. In all the examples, as usual in this book, the investments will be in fixed-income securities. However, the concepts could apply to other situations, such as investments in businesses, new projects, real estate, or stocks.

SOURCES OF RETURN

When you set up and manage such a fund, you will have returns from three sources:

1. The interest income you earn on your original investment
2. The interest income you earn by investing the interest of your original investment. This is called "interest on interest"
3. The capital gain or loss when the fund is closed out or when bonds mature or are sold during the life of the fund or on the closeout date

HOW TO ANALYZE THIS PROBLEM

To analyze this problem properly, you must make some decisions about the fund you are setting up. You must choose

1. The initial contribution
2. The interest rate earned on the reinvested interest
3. The time when the fund will be dissolved, called the investment horizon
4. The value of the investments in the fund when the fund is dissolved

Each of these decisions will partly determine the total amount in the fund when the fund is dissolved.

The interest rate (nominal annual rate compounded semiannually) that equates the present value of the final amount with the amount originally invested is called the rate of total return. The final, or total, amount in the fund includes the interest earned on the original investments, the interest earned on the reinvested interest, and the capital gain or loss when the investments mature or are sold at the end of the horizon.

Here are three examples of how this total return fund works.

EXAMPLE 13.1.   You have an 8% bond, bought at 100 and maturing in 5 years; an investment horizon of 5 years; and an assumed reinvestment rate of 8%. You have a coupon payment of 4 every 6 months for 5 years, for a total of 10 payments. From the table for the future value of an annuity, with n = 10 and i = 4, we have the future value of the annuity = 12.0061071. Therefore,

The total value of coupons reinvested = 4 × (12.0061071) = 48.024428

We compute the future value of the investment as follows:

Interest payments	40.000
Interest on interest	8.024428
Maturity value of bond	100.000
Total	148.024428

The present value of investment is the amount originally invested, or 100.000. Then the total return will be y in Equation 13.1.

$$100\left(1+\frac{y}{2}\right)^{10} = 148.02448 \qquad\qquad \text{Equation 13.1}$$

Solving this equation, we find that y = 8%. This is what we would intuitively expect, with an 8% coupon and reinvestment rate, a purchase at par, and maturity at par.   ∎

EXAMPLE 13.2.   We now still have a 5-year 8% bond bought at 100, but now we have a reinvestment rate of only 5%. The rest of the example is the same as Example 13.1.

In this example, we have a lower reinvestment rate, so we expect a somewhat lower total return. However, the coupon income will still be at 8%, so the total return should not be a great deal lower than 8%.

We have a coupon payment of 4 every 6 months for 5 years, for a total of 10 payments, invested at a 5% annual rate, compounded semiannually. This gives us the future value of an annuity of 1, using n = 10, i = 2.5, and = 11.2033818. Then

The total value of coupons reinvested = 4 × (11.2033818) = 44.813527

The future value of total investment will be as follows:

Interest payments	40.000
Interest on interest	4.813527
Maturity value of bond	100.000
Total	144.813527
Present value of investment	100.000

This will give a total return = 7.54% per year, compounded semiannually. ∎

EXAMPLE 13.3.   In this example, we still have an 8% bond bought at 100, a 5-year horizon, and a reinvestment rate of 5%. However, we now assume that the bond is worth 110 at the end of 5 years; note that no statement is made in this case about the final maturity date of the bond. We would expect an increase in total return from Example 13.2, but it is not immediately clear whether we would also have an increase from Example 13.1. The capital gain will offset the lower reinvestment rate, but the extent of the offset is not immediately clear.

We still have a coupon payment of 4 every 6 months for 5 years, for a total of 10 payments. The future value of the annuity is the same as in Example 13.2, for n = 10, i = 2.5, and = 11.2033818. Therefore,

The total value of coupons reinvested = 4 × (11.2033818) = 44.813527

For the future value of total investment, we have the following:

Interest payments	40.000
Interest on interest	4.813527
Capital appreciation	10.000
Par value of bond	100.000
Total	154.813527

As before, the present value of the investment = 100.000. We calculate the total return = 8.93% per year, compounded semiannually.   ∎

EXAMPLE 13.4.   In this case, we have a bond maturing in 30 years, bought at 100, with a coupon and yield of 5.25%, about the yield of the long-term Treasury in the fall of 2001. We assume the interest earnings will be reinvested at the original investment rate of 5.25%.

The future value of an annuity of 1, 30 years, 5.25% compounded semiannually = 142.231821. Then the future value of the fund will be as follows:

Interest payments	157.50
Interest on interest	215.85853
Par amount of bond	100.000
Total	473.35853

The total return is 5.25%, as expected.   ■

EXAMPLE 13.5.   This is the same as Example 13.4, except that we assume that the bond has a value of 110 in 30 years. (Of course, the bond will now have a longer maturity period than 30 years.)

Then the future value of the fund will be as follows:

Interest payments	157.50
Interest on interest	215.85853
Par amount of bond	100.000
Capital gain	10.000
Total	483.35853

This produces a total return of 5.32%.   ■

SOME OBSERVATIONS ON THESE EXAMPLES

You can see that, although you must make assumptions about the reinvestment interest rate and the value of the bonds at the end of the horizon, these assumptions will have different importance in determining the total return. In Example 13.1, the interest on interest amounted to only about 20% of the coupon income. In Example 13.4, the interest on interest was more than 37% higher than the coupon income. Generally, the reinvestment rate becomes increasingly important as the length of the time horizon increases.

In Examples 13.3 and 13.5, the maturity amount was increased by 10 points from Examples 13.2 and 13.4, respectively. For Examples 13.2 and 13.3, this increase resulted in an increase of about 1.39 percentage points in the total return. For Examples 13.4 and 13.5, a similar point increase resulted in an increase of only about .07 percentage points in the total return, a very much smaller amount. Generally, the capital gain or loss becomes less important as the time horizon increases.

EXTENSIONS OF THE ANALYSIS

The above examples assumed that (1) no investments or withdrawals were made into or from the fund, (2) that the interest reinvestment rate was constant, and (3) that no gains or losses occurred until the fund was closed out. However, you could vary the reinvestment rate, allow for additional contributions or withdrawals, and allow for gains or losses. Computer programs, possibly including some spreadsheet programs, could allow you to do this fairly easily. For example, you might believe that interest rates will increase for the next 10 years, reaching a limit. You could program this increase into your analysis to reach a result more in line with your predictions.

CHAPTER SUMMARY

An investment fund may have earnings from three sources:

> Interest earned
> Interest on interest
> Capital gain or loss when bonds mature or are sold, or the fund is
> dissolved

Analysis requires choosing:
> Initial contribution
> Interest rate earned on reinvested interest
> Investment horizon, time when the fund will be dissolved, or the analysis
> stops
> Value of the fund's investments upon dissolution

Total rate of return is the rate that makes the present value of the fund's final value equal to the initial contribution.

As the investment horizon increases, reinvestment rate generally becomes increasingly more important, and capital gain or loss becomes increasingly less important.

COMPUTER PROJECTS

1. Use your favorite mathematical program to develop a display showing the total return for various combinations of coupon rate, purchase price, interest reinvestment rate, time horizon, and final price. Display the total return for various combinations.
2. Now vary the interest reinvestment rate during the time period of the fund. What results do these changes have in the total return?
3. Now keep the reinvestment rate constant, but vary the final maturity value of the holdings. What results do these changes have in the total return?
4. Build on Project 3 by varying the investment horizon. What results do these changes have in the total return? What conclusions do you draw?

TOPICS FOR CLASS DISCUSSION

1. What problems would you have in actually implementing a total return model? How would you overcome these problems, or at least try to? Where might the main problems lie for the models you are likely to set up?

PROBLEMS

1. You have placed $10,000 into a special account for your son's college education. You will need the money in 15 years. You have bought a 6% bond at par, which matures in 15 years, but you believe that interest rates will rise and you can invest the interest earned at 7%. How much will you have in 15 years?
2. You are 35 years old and plan to retire at age 65. You are setting up a retirement account. What is your most likely investment horizon?

CHAPTER 14

Volatility and Its Measures

This chapter introduces the concepts underlying the interest in duration and convexity. It covers volatility, why it is important, and how we measure it.

When you finish this chapter, you should understand the importance of volatility, the characteristics of volatility, and the three ways we measure it.

WHAT IS VOLATILITY?

Volatility is the relative tendency of a bond price to change when the yield changes. Some bonds can clearly change more than others for a particular change of yield. For example, if a 30-year 6% bond moves from a 6% yield (a price of 100) to a 6.25% yield, the price will decline by 3.37% to a price of 96.63. A T-bill maturing in 1 week, on the other hand, will decline from a price of 99.883 to 99.878, a decline of about .005%. The 30-year bond is much more volatile than the T-bill maturing in 1 week.

WHY DO WE CARE ABOUT VOLATILITY?

Volatility can measure one kind of bond risk, or portfolio risk. By measuring risk, we can determine the risk of a bond or of a portfolio. This also means that risk is not something to be avoided; indeed, one cannot avoid risk completely in any investment. But the portfolio manager who can *measure* portfolio risk can also *manage* the portfolio's risk. The portfolio manager selects investments so as to maintain a predetermined level of risk in the portfolio.

Volatility has become much more important in bond portfolio management than it was years ago. Partly this is due to increased analytical knowledge about bond management, but also it is because bond prices are more volatile than they were years ago. This has given volatility a much greater importance than it previously had, and this volatility has also resulted in numerous articles in the press. Nowadays, bonds may fluctuate more in one day than in a whole year many years ago.

For example, on the day this discussion were written, the United States Treasury announced that is was suspending the issuance of 30-year bonds. As a result, the 30-year bond rose more than 5 points, the largest rise in 13 years. The bond rose more than 2 points the following day, but gave up most of that additional gain on the next trading day.

No one really knows the reason for the increase in bond market volatility during the past few decades. Inflation, increased bond trading, increased bond investment, and the increasing use of bond-related derivatives may all play a part. But bonds do seem much more volatile than they were a few decades ago. This volatility shows no sign of going away, so it seems to be a permanent fixture of the bond market.

WHAT VOLATILITY MEASURES CAN DO FOR YOU

Volatility measures let you predict how a bond will change in price as the market yields change. Alternatively, they will tell you what must happen to yields in order for a given price change to occur. You can apply them to individual bonds, and you can apply some of them to an entire portfolio.

Using volatility measures, you can measure risk and return possibilities and measure the risk or chance of gain or loss. You can measure profit or loss potential.

You can set and implement investment policies and strategies in terms of desired volatility measures. You can compare investments and analyze swaps.

MEASURES OF VOLATILITY

We have three measures of volatility:

1. Price change per unit change of yield
2. Duration
3. Convexity

We will discuss duration and convexity in the next three chapters.

PROPERTIES OF VOLATILITY

VOLATILITY IS NOT THE SAME FOR ALL BONDS

At first, this looks like a truism, and it is to some extent. However, it also means that each bond has a unique, measurable volatility, which can be used for analytical purposes.

PRICE VOLATILITY IS APPROXIMATELY SYMMETRIC FOR SMALL YIELD CHANGES

You saw, from the price-yield curve diagram in Chapter 6 that this curve is generally convex. However, in a small interval around any particular yield, a straight line, the tangent line, is a good approximation for the actual curve. The tangent line is symmetric, so that the resulting approximation of price is also symmetric.

For example, suppose we have a 4%, 30-year bond at par (100), and impose a 10 basis point change. What happens to the price? If we increase the yield by 10 basis points (to 4.1%), the price decreases 1.72%. If we decrease the yield by 10 basis points (to 3.9%), the price increases 1.76%. You can see that the price change is approximately symmetric.

PRICE VOLATILITY IS NOT SYMMETRIC FOR LARGE YIELD CHANGES

Suppose we have the same 4%, 30-year bond at par, but have a 300 basis point change. If we increase the yield by 300 basis points (to 7%), we decrease the price by 37.42%. If we decrease the yield to 1%, we increase the price by 77.59%. You can see that the price changes are not even approximately symmetric.

PRICE INCREASES ARE GREATER THAN
PRICE DECREASES

Price increases due to yield declines are, generally, greater than price decreases do to the same size yield increases. This is due to the convexity of the price-yield curve and is shown in the previous example. In that admittedly extreme case, the increase was about twice the decrease for the same basis point change.

VOLATILITY INCREASES AS MATURITY INCREASES

Example. If a 1-year 4% bond at a 4% yield has a 10 basis point change in yield, the change is .1% in price, for either a yield increase or decrease. If a 30-year bond has the same 10 basis point change, the change is 1.76% for a yield decrease or a 1.72% change for a yield increase.

VOLATILITY DECREASES AS COUPON INCREASES
(LOWER COUPON BONDS ARE MORE VOLATILE)

For example, if a 4% 30-year bond at 4% yield changes 10 basis points, the price changes 1.76% or 1.72%, depending on whether the change is a decrease or an increase. A 6% 30-year bond at a 4% yield changes 1.62% or 1.58%, depending on whether the yield decreases or increases. The higher coupon has a lower percentage change.

VOLATILITY INCREASES PER UNIT OF YIELD
AS YIELD DECREASES (LOWER YIELDS ARE
MORE VOLATILE)

Suppose we have a 30-year 4% bond. If the yield moves from 8% to 8.1%, the price changes from 54.75 to 54.06, down 1.26%. If the yield moves from 2% to 2.1%, the price changes from 144.96 to 142.13, down 1.95%.

VOLATILITY DECREASES AS BONDS SELL SUBJECT
TO THE CALL FEATURE

Suppose a 10% bond may be called now at par (100). For yields much under 10%, say less than 8%, wide fluctuation in yields, say down even to under 1%, will result in no price increase. Buyers simply will not pay more than par for a bond that could be called at par at any time. In general, bonds that may be called in the near future won't rise much over the call price because the issuer may call the bonds soon.

Most municipal bonds, many corporate bonds, and even some Treasury bonds have call features. If you do any bond investment, you should know the call features of your bond investments.

BOND INVESTMENT MANAGEMENT INTERLUDE

In the 1949 edition of his famous book *The Intelligent Investor*, Benjamin Graham comments on the price change of the Atchison Railroad General 4s of 95. Graham states that these bonds fell from a high price of about 141 in 1946 to 115 within 2 years, and he suggests that the purchase of long-term, high-quality bonds at historically high premiums is not good business. This corresponds with the statement made earlier that low yields are more volatile.

MEASURING VOLATILITY BY MEASURING PRICE CHANGE PER UNIT CHANGE IN YIELD

This is the easiest of the three measures to understand. The volatility measure is simply the direct percentage amount the price changes as the yield changes. Note that the yield change per unit change in price is the reverse of this process. We looked at the price-to-yield calculation in Chapter 12.

This concept is easy to understand and, using the equations in Chapter 6, can be computed. It directly measures performance and can be applied to portfolios for portfolio measurement. It can apply to the entire portfolio or to segments of a portfolio.

However, you cannot use mathematical analytical techniques with this method as easily as you can by using duration and convexity. As a result, for more detailed analyses, most people use the other measures, which we will discuss these in the next three chapters.

CHAPTER SUMMARY

Volatility is the relative tendency of a bond price to change when yield changes.

Volatility measures can measure bond risk.

Measures of volatility include price changes when yield changes, duration, and convexity.

TOPIC FOR CLASS DISCUSSION

Examine price records of high-quality, long-term bonds in times of bond bear markets, such as from 1946 to 1980. How did the high-premium bonds fare during these bear markets? What might you conclude from this effect? How would you manage the eight volatility properties listed in the chapter when managing your own portfolio?

CHAPTER 15

Duration

This chapter and the next two chapters discuss duration and convexity. We develop duration as a time measure and also as a measure of risk. We take a historical approach by developing a set of measures for this length of time and presenting duration as the natural extension of a series of such measures. All these measures have some use in finance. You should know about them, how to compute them, and when some of them might be used.

Chapter 16 examines convexity, once more in a traditional way, as an equation involving the sum of future payments, present values, time periods, and constants. We use convexity to compute an additional adjustment in the predicted bond price for a given yield change.

Chapter 17 presents a calculus derivation of duration and convexity and uses a Taylor's Series expansion to demonstrate the use of duration and convexity for estimations of percentage change in bond price, given a change in yield. This chapter presents a different development of duration and convexity, which may make more sense to a person who has studied calculus. This approach leads immediately to additional ideas on mathematical applications in finance. These ideas include negative duration and negative convexity.

If you have studied calculus, read all three chapters. They will give you an insight into the historical development of an important concept of fixed-income mathematics and an intuitive idea of where they might be used, as well as the mathematical background for further use. If you have not studied calculus, you should at least read the first two of the three chapters. Perhaps you should also study calculus.

Duration is a risk measure as well as a time measure. We present this discussion of duration as a natural outgrowth of our presentation of risk (or volatility) measures in Chapter 14. We show how duration displays the same requirements of risk measures that we discussed in the previous chapter, and so it is a natural risk measure for analytical purposes.

When you finish this chapter, you should understand the concept of duration as an extension of time measures of a flow of funds. You should know how to compute the duration of a flow of funds. You should understand portfolio duration and how to compute it.

HISTORICAL BACKGROUND

Duration was first invented by Frederick Macaulay in the 1930s as part of his analytical work on American corporate bonds. He published his study in 1938. His work over the years did not attract much attention and was eventually forgotten. Since Macaulay, a number of eminent economists independently reinvented duration, but Macaulay invented it first, so he gets the credit.

During the 1970s, developments in finance caused a rediscovery of Macaulay's work, and it became widely used in portfolio theory. Later, the idea of duration was expanded to include convexity as a portfolio management tool. We cover convexity in the next chapter.

HOW LONG WILL A FLOW OF FUNDS
BE OUTSTANDING?

Suppose you want to do some analytical work studying various flows of funds, and you want a measure of how long a particular flow of funds will be outstanding. Table 15.1 presents an example of the problem.

We assume that payments are annual, for simplicity, and that the payments of 101 and 110 include an interest payment of 1 and 10, respectively, and a maturity payment of 100 in both cases. The other payments are interest only.

In this case, it is not immediately intuitively clear which flow of funds will be outstanding longer, if we include the values of the interest payments. The last payment in Cash Flow 1 occurs 1 year earlier than that in Cash Flow 2,

TABLE 15.1   Cash Flow Examples

Year	Cash flow 1	Cash flow 2
1	1	10
2	1	10
3	1	10
4	101	10
5	0	110

but Cash Flow 2 has bigger interest payments, which would tend to shorten the outstanding time, at least in some measures. Here are some ways to measure the length of time a cash flow will be outstanding:

TIME TO MATURITY

This measure uses time to maturity as the time the flow of funds will be outstanding. In this case, Cash Flow 2 has the longer time measure. It is outstanding 5 years, and Cash Flow 1 is outstanding only 4 years, because their final maturities are in 5 and 4 years, respectively. This measure is not quite as strange as it might at first appear. Time to maturity is the duration for a zero-coupon bond. You can easily understand the concept of time to maturity, it is not related to the coupon rate of the bond, and no computation is involved.

AVERAGE LIFE

Average life is the sum of the individual maturities in the flow, multiplied by the time to maturity, all divided by the total amount of the maturities. The equation is follows:

$$\text{Average life} = \frac{\sum_i t_i M_i}{\sum_i M_i} \qquad \text{Equation 15.1}$$

where M_i are the maturities at time t_i.

Suppose you have a portfolio of $5,000 in bonds maturing in 3 years, $8,000 maturing in 4 years, and $7,000 maturing in 6 years. What is the average life of your portfolio? Table 15.2 shows how to compute the average life of the portfolio.

In the two examples, Cash Flows 1 and 2, each cash flow has only one

TABLE 15.2 Average Life Calculation Example

Time	Amount	Time × Amount
3	5,000	15,000
4	8,000	32,000
6	7,000	42,000
Totals	20,000	89,000

Average life = (89,000)/(20,000) = 4.45 years

maturity, so Cash Flow 1 has an average life of 4 years, and Cash Flow 2 has an average life of 5 years.

Average life can apply to a portfolio and does not require coupon rates. You can use it for analytical purposes. It has been used for many years in the municipal bond business, in new issue bond bidding scale calculations, and in pricing bonds with a mandatory sinking fund. The sinking fund bond is priced to the average life date (month, day, and year). Average life does not contain a present value factor.

WEIGHTED AVERAGE CASH FLOW

Weighted average cash flow expands from average life to include the coupon payments. The equation is as follows:

$$\text{Weighted average cash flow} = \frac{\sum_i t_i(c_i + M_i)}{\sum_i (c_i + M_i)} \qquad \text{Equation 15.2}$$

where c_i, M_i are the coupons and maturities payable at time t_i.

Weighted average cash flow is considered superior to average life because it includes coupon payments. However, this also requires that the coupons either be already known or estimated. Like average life, weighted average cash flow also does not contain a present value factor.

In the previous example, Cash Flow 1 has a weighted average cash flow of 3.94 years, and Cash Flow 2 has a weighted average cash flow of 4.33 years.

DURATION (MACAULAY DURATION)

Macaulay duration continues the previous development in a natural way by adding a factor for present values.

$$\text{Macaulay duration} = \frac{\sum_i \left(\dfrac{1}{1+\dfrac{y}{2}}\right)^i t_i (c_i + M_i)}{2\sum_i \left(\dfrac{1}{1+\dfrac{y}{2}}\right)^i (c_i + M_i)} \qquad \text{Equation 15.3}$$

$$= \frac{\sum t \bullet PVCF_t}{k \bullet PVTCF} \qquad \text{Equation 15.4}$$

where $PVCF_t$ = Present value of the cash flow at time t, using the yield to maturity (or IRR) of the security

$PVTCF$ = Present value of the total cash flow, which also
= Bond price plus accrued interest

t = Number of periods to cash flow payment (half years, in the case of Equation 15.3)

k = Number of payments per year ($k = 2$ in the United States)

y = Yield used to compute bond price

c_i = Coupon payment at time i

M_i = Maturity payment at time i

Equations 15.3 and 15.4 extend the equation for weighted average cash flow (Equation 15.2) by including a present value factor in both numerator and denominator. This means that the yield must be known or estimated to compute the duration. Changing the yield changes the duration.

The denominator in Equation 15.4 shows the present value of the total flow of payments. This includes both the bond price and the accrued interest. Twice per year, on the interest payment dates, this equals the bond price, because the accrued interest equals 0. These semiannual dates are also the dates for which one is likely to compute duration for an individual bond. However, always remember that the denominator includes accrued interest as well as the bond price. This also means that the denominator is not continuous, but rather saw-toothed in shape. It goes down by the amount of an interest payment on each interest payment date. You can also think of it as the "dirty price."

Observe also that the duration equation can apply to a portfolio of many bonds, or even to any flow of funds. The yield used to calculate the duration would be the IRR that equates the present value of the outflows and the inflows. For a portfolio, the "outflows" would be the market value of the portfolio at time $t = 0$.

We use duration directly as a measure of bond price volatility. It satisfies the functional relationships required of this measure: longer maturities have longer duration, and lower coupon bonds have longer duration, other factors being equal. However, duration is a complicated formula and not easily intuitively understood. With practice, though, you can achieve some intuitive understanding of it and estimate the duration of many cash flows.

Observe that the choice of time periods must be consistent with the choice of i in Equations 15.3 and 15.4. For semiannual cash flows, if you start with i = 1 (half-year periods), then the resulting duration will be measured in half years. Divide by 2 (k = 2) to get the duration in years, the usual measure for duration. Equation 15.3 assumes half-year periods. Equation 15.4 is more general.

MODIFIED DURATION

We use duration as a measure of bond risk. Its importance is linked to bond volatility. Modified duration is a better measure of bond volatility, or bond risk.

For most bond analytical work, we need to know the percent price change, rather than the dollar price change, because return is expressed as a percent. The equation for an approximate percent price change is as follows:

Approximate percentage price change =

$$-\left(\frac{1}{1+\frac{y}{2}}\right) \cdot \text{Macauley duration} \cdot \text{Yield change} \qquad \text{Equation 15.5}$$

We define modified duration as follows:

$$\text{Modified duration} = \left(\frac{1}{1+\frac{y}{2}}\right) \cdot \text{Macauley duration} \qquad \text{Equation 15.6}$$

Then we have

$$\begin{aligned}&\text{Approximate percentage price change} \\ &= -(\text{modified duration})(\text{yield change})\end{aligned} \qquad \text{Equation 15.7}$$

PAYBACK INTERLUDE

Here is another time measure, which isn't related to any of the previous measures but is widely used, especially in manufacturing. It is called payback, and it is the time required to recover the investment amount. You should know about payback, how to compute it, and its advantages and drawbacks.

Here is an example of payback. Suppose you run a manufacturing operation, and you make some improvements to your production line (at a cost

$10,000) that save you $2,000 per month. Payback is 5 months, the time required to recover your $10,000 investment at the rate of $2,000 per month. No present value calculations are involved at all. Suppose a different investment of $10,000 would save you $2,500 per month. This has a shorter payback, so it is the preferred investment. You get your money back in 4 months instead of in 5 months.

But payback also has drawbacks. Suppose you have two choices for investing $10,000. Choice 1 repays $2,500 monthly, and Choice 2 repays $2,000 monthly. Choice 1 has the shorter payback, so you choose it. But suppose Choice 1 only pays for 6 months ($15,000 total), and Choice 2 pays for 15 months ($30,000 total). You don't need to do an elaborate present value analysis to realize that Choice 2 is better, even though the monthly earnings are smaller, because these earnings continue for a much longer time. For most cases where payback is actually used, using present values wouldn't change the final decision on whether or not to do the project.

A PLEA FOR PAYBACK

Payback is still, apparently, widely used in U.S. manufacturing operations. The United States is well represented in international markets, at least for certain manufactured goods. The men and women who run this country's manufacturing plants are exceedingly competent at their jobs. They must be doing something right, and they like payback.

Payback is easy to use and to understand. You can easily make informal adjustments, based on your own ideas of how the flow of funds might actually work out. It leads to decisions that usually are not egregiously wrong. It can easily be explained to the plant workforce and to the management. Payback has many advantages.

I have not heard of much use of payback in finance, although I have heard that it is sometimes used in emerging country markets. You might expect a country to default in 5 years, but if you are repaid in two years, you might not much care.

The payback concept is probably much more widely used, at least informally, for both personal and business analyses, than is generally realized.

CALCULATION OF MACAULAY DURATION AND MODIFIED DURATION

Here is an example of the calculation of both Macaulay duration and modified duration. We'll use the 4% bond due in 3 years at a 4% yield, as shown earlier in the book.

First, approximately what size do we expect the duration to be? Duration is a time measure, and it measures the time a series of payments will be outstanding. Our bond matures in 3 years, and if the coupons were zero, the stream would have only one payment in 3 years. That would make the duration 3 years. We have coupon payments, which would tend to shorten the duration. Increasing coupon size decreases duration, and we are increasing the coupon from zero to 4% annually. This will shorten the duration somewhat. However, the coupon is still not very large, and the bond matures reasonably soon, so we don't expect a large reduction in duration. Therefore, we expect the duration to be a "slightly" less than 3 years. See Table 15.3.

The Macaulay duration is 2.856 years, and the modified duration is 2.801 years. Both are slightly below the 3-year duration of the zero-coupon bonds and are within the expected reasonable range of "slightly less than 3 years." An answer greater than 3 would be clearly wrong. An answer much less than 2 would also likely be clearly wrong. It is frequently wise to do checks for reasonableness before you use the results of your calculations. This is one reason that you shouldn't simply accept the results your computer or calculator gives you without at least some informal check for reasonableness. It also means that you should know what you are doing, rather than simply enter data into a computer or calculator.

In the calculation of Macaulay duration in half years, we divided by 100 because it is the price of the bond, not because one divides by 100 (a mistake

TABLE 15.3   Calculation of Macauley Duration and Modified Duration (4% bond due in 3 years at a 4% yield)

Period	Cash flow	Present value of 1 @ 2%	PVCF$_t$	$t \times$ PVCF$_t$
1	2.00	.980392	1.960784	1.960784
2	2.00	.961169	1.922338	3.844676
3	2.00	.942322	1.884644	5.653932
4	2.00	.923845	1.847690	7.390760
5	2.00	.905731	1.811462	9.057310
6	102.00	.887971	90.573080	543.438540
			100.000000	571.346002

$$\text{Macauley duration} = \frac{571.346002}{100} = 5.71346$$
(in half years)

$$\text{Macauley duration} = \frac{5.71346}{2} = 2.856$$
(in years)

$$\text{Modified duration} = \frac{2.856}{1.02} = 2.801$$

frequently made by students in my classes). We compute the Macaulay duration in half years because the periods are numbered in half years. We divide
by 2 to get the duration in years; this "2" is the "k" in the denominator of the
duration equation. If we had numbered the periods in full years—that is, from
0.5 to 3.0—the calculation would have produced Macaulay duration in years,
and no division by "2" or "k" would have been needed. We divided by 1.02
to compute the modified duration because y = 4%, so y/2 = 2% = .02.

USING MODIFIED DURATION TO PREDICT PRICE CHANGE

We now use the modified duration computed above to calculate the expected
new price for changes in the yield. We will compute the new price for yield
changes of 10 basis points (0.10%) and 300 basis points (3.0%) for our 4%
bond due in 3 years.

EXAMPLE 15.1.   Use the following duration calculations to compute the
10 basis point change from 4.00 to 3.90 and 4.10:

$$(2.801) \times (.0010) = .002801 = .2801\%$$

	Old price	New price (estimate)	Actual	Delta
Decrease yield	100.000	100.280	100.28	0.00
Increase yield	100.000	99.720	99.72	0.00 ∎

EXAMPLE 15.2.   Use this calculation to compute the 300 basis point
change from 4.00 to 1.00 and 7.00:

$$(2.801) \times (.03) = .08403 = 8.403\%$$

Decrease yield	100.000	108.403	108.84	0.437
Increase yield	100.000	91.597	92.01	0.413

You can see that using the modified duration along for a yield change of 10
basis points has predicted very closely the correct price. However, for a 300
basis point change, the price is off by almost .5%. The predicted price is lower
than the correct price. The price predicted by the duration adjustment generally somewhat underestimates the correct price. We'll correct this (mostly) in
the next chapter, using the adjustment for convexity.

Note also that the duration prediction is symmetric. The predicted price
adjustments up (for decreasing yield) and down (for increasing yield) are the
same.

Remember that the adjustment predicted by using modified duration is the
percentage change in price, not the dollar change in price. In the preceding

example, the dollar price was 100, so the percentage price change and dollar price change were the same. Usually, this won't be the case.   ■

DOLLAR DURATION

Sometimes people want to compute an actual price change, rather than a percentage price change. To do this, use dollar duration.

> Approximate dollar price change
> = (Approximate percentage price change)(Initial price)
> = (Modified duration)(Yield change)(Initial price)

> Dollar duration = (Modified duration)(Initial price)

Therefore,

> Approximate dollar price change
> = (Dollar duration)(Yield change)

A PICTORIAL VIEW OF DURATION

Many students learn from diagrams. Figure 15.1 displays a useful diagram to show the duration. Imagine a seesaw, resting on a fulcrum. The seesaw has a

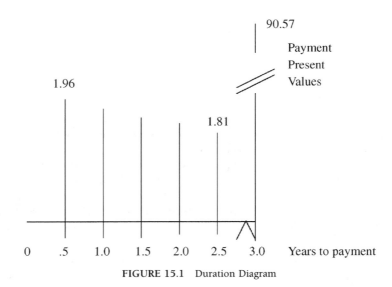

FIGURE 15.1   Duration Diagram

series of time markings on it. Each time marking has a weight of the present value of the payment due at that time. The time marking that balances the seesaw is the duration.

Here is a seesaw chart for the 4% 3-year bond at a 4% yield. We use the values shown in Table 15.3. The fulcrum is at 2.86 years.

Proof. The diagram in Figure 15.1 may not be immediately obvious. Here is the proof.

Let t_i = Time payment P_i is made
Let P_i = Payment made at time t_i
Let D = Balance point of the seesaw, measured in time
Let y = Interest rate of the evaluation

Then, at D, the downward force of the right-hand side (RHS) equals the downward force of the left-hand side (LHS) around D.

$$\text{Downward force (LHS)} = \sum_i (D - t_i) \bullet P_i \bullet \left(\frac{1}{1+y}\right)^{t_i}, t_i \leq D$$

$$\text{Downward force (RHS)} = \sum_i (t_i - D) \bullet P_i \bullet \left(\frac{1}{1+y}\right)^{t_i}, t_i > D$$

Then

$$\sum_i (D - t_i) \bullet P_i \bullet \left(\frac{1}{1+y}\right)^{t_i} (t_i \leq D) = \sum_i (t_i - D) \bullet P_i \bullet \left(\frac{1}{1+y}\right)^{t_i} (t_i > D)$$

And

$$\sum_i (D - t_i) \bullet P_i \bullet \left(\frac{1}{1+y}\right)^{t_i} = 0, \text{for all } t_i$$

Then

$$D \sum_i P_i \bullet \left(\frac{1}{1+y}\right)^{t_i} = \sum_i t_i \bullet P_i \bullet \left(\frac{1}{1+y}\right)^{t_i}$$

And

$$D = \frac{\sum_i t_i \bullet P_i \bullet \left(\dfrac{1}{1+y}\right)^{t_i}}{\sum_i P_i \bullet \left(\dfrac{1}{1+y}\right)^{t_i}}$$

which is the definition of Macaulay duration. For teaching purposes, we have made the compounding period for y equal to the time periods. ∎

A MISCONCEPTION ABOUT DURATION

I have occasionally heard that duration is "when you get half your money back." This is an incorrect view of the meaning of duration. For example, in the case of a zero-coupon bond, the duration is the time to maturity. At the duration time, you get all your money, not just half. In the previous example, where you had a 4% bond, due in 3 years, at a 4% yield, you had not even received 10% of the total money at the duration time. You will receive most of it when the bonds mature, shortly after the duration time.

PORTFOLIO DURATION

A portfolio is a collection of bonds, considered as a group. Any group of bonds could be considered a portfolio, but most portfolios have some feature or features that distinguish them from other portfolios. These features describe the characteristics of the portfolio. Usually, these include the purpose of the portfolio, which would in turn determine the types of bonds bought by the portfolio. Here are some examples of portfolios frequently encountered:

1. You have invested your $200,000 inheritance in 10 annual maturities, going out 10 years, of $20,000 each in Treasury securities. This is called a "laddered" portfolio. A laddered portfolio is a standard investment technique used by many individual investors with smaller portfolios. In bond investment management, a portfolio of that size would be considered extremely small, even though it is extremely large for most people.

2. You are the chief financial officer of a large corporation. You might have a portfolio of Treasury bills to pay near term expenses, a portfolio of medium-term bonds to invest the company's surplus funds, and a portfolio of bonds specifically set aside to finance a planned new construction project. Each of these portfolios has a different purpose, each might have different management techniques, and each might be quite different from the others. The precise investments would depend on the purposes. For example, if your construc-

tion project would cost $100,000,000 and would be finished in 2 years, you might have 20 monthly maturities of $5,000,000 each to pay monthly bills as they come in.

3. You run a family of mutual funds. You would probably have one or more of at least the following: short-term money market funds, both taxable and tax-exempt; medium- and long-term municipal bond funds; a mortgage-backed securities fund; and medium-term and long-term Treasury funds.

4. You are the treasurer of a village in Westchester County, New York. You receive almost all your tax revenues on or about July 31. You have a budget for the coming year, which includes scheduled payments for payroll, debt service, snow removal, and other expenses, going to next June 30. You invest in a portfolio of short-term Treasuries with maturities corresponding to the spending requirements of your budget.

COMPUTING PORTFOLIO DURATION

Here are two methods of computing the modified duration of a portfolio. Both methods require knowledge of the market value of the individual bonds in the portfolio and the total market value of the portfolio.

Method 1. The portfolio's modified duration is approximately equal to the weighted average of the modified durations of the individual bonds in the portfolio.

Let b_i = Bond i in the portfolio
w_i = Weight of bond i in the portfolio, which also
= (Market value of b_i)/(Market value of portfolio)
d_i = Modified duration of b_i

Then

$$\text{Portfolio modified duration} = \sum w_i d_i$$

Method 2. The portfolio's modified duration is the modified duration of the entire schedule of interest and principal payments at the yield equal to the internal rate of return (IRR). The IRR is the yield that sets the present value of the portfolio's flow of payments, interest, and principal equal to the market value of the portfolio plus the portfolio's accrued interest. Compute the IRR (= y) using a standard numerical analysis technique, such as the bisection method discussed earlier. Then use the resulting yield (y or $y/2$) to compute the modified duration, using the equations shown earlier in this chapter.

Portfolio modified duration is a statistical measure. You should treat it with the same respect that you treat any statistical measure. It is meant to describe a set of data. In the case of portfolios, the duration of a heterogeneous set of bonds might have relatively little meaning, although it could be computed. For portfolio duration to have a meaning, the bonds should be related in some way and have enough features in common for this measure to make sense. Generally, the bonds should perhaps have close individual durations, close maturity dates, and change by approximately the same basis point amount in order for portfolio modified duration to have a meaning.

USING DURATION AS A PORTFOLIO MANAGEMENT TOOL

Duration measures the risk of the portfolio. One way to manage a portfolio is to select a duration; this measures the portfolio's risk. The management sets the portfolio management objectives, at least partly in terms of duration. It can then measure the portfolio manager's performance in meeting the risk requirements of the portfolio. A manager sets up the portfolio investments to achieve the duration objective, as well as other portfolio objectives and constraints, such as the issuer, credit rating, maturity, and tax-exemption features.

For some portfolios, this is not really necessary. Other constraints will control the portfolio's content. For example, in the portfolio for the village and the corporation's portfolio to pay current bills, the term is short and the duration is not really important. But for the portfolios of the mutual fund family, the durations of the various portfolios could be quite important. In the case of your laddered portfolio, the duration is related to the 10-year period you chose for the ladder. In doing this, you set a risk level you could accept. You could just as well have measured this risk level in terms of duration. In this case, you managed the portfolio by choosing a risk level (which could be stated in terms of duration), the quality (U.S. Treasuries), and the ladder concept.

DURATION FOR BONDS WITH EMBEDDED OPTIONS

Suppose your bond has a call feature. This kind of feature is called an embedded option. As the yield changes, the embedded option might start to affect the price. For example, if the yield becomes lower than the coupon rate, and the bond can be called at par, the bond will be priced to the option date and call price. This will change the duration, in this particular case.

But a more complete analysis will consider the chance of the bond being called. This leads to the concept of effective duration, which involves use of probabilities. We'll cover effective duration later in the book, in one of the chapters on probability applications.

NEGATIVE DURATION

Can duration ever be negative? Yes. In Chapter 17 on the calculus derivation of duration and convexity, we'll discuss negative duration and convexity and what they mean.

PROBLEMS WITH DURATION AS A MEASURE

Duration has some problems as a measure of risk, or sensitivity. You should at least be aware of these problems.

Duration does a good estimating job only for small changes in yield. The predicted price's difference from the actual price increases as the yield change increases, due to the generally convex nature of the price yield curve.

Portfolio duration is a good predictor only if all the bonds in the portfolio change by approximately the same basis point amount.

Duration is inadequate for bonds denominated in multiple currencies, because the currencies themselves fluctuate in value, and the fluctuations affect the bond prices.

Duration may not be appropriate for cash flows that are sensitive to changes in interest rates. These include bonds with embedded options, such as a call option, which may be exercised at any time, and especially include mortgage backed securities (MBS). These securities require careful analysis if you want to use the duration concept. We'll discuss MBS in Chapter 21.

CHAPTER SUMMARY

We started with duration as a measure of the length of time a bond (or a series of cash flows) was outstanding.

Duration measures both time and risk, in our presentation.

Duration's importance is as a measure of bond volatility.

Macaulay duration is a measure of the weighted average life of bond, including coupons and a present value factor.

$$\text{Modified duration} = \left(\frac{1}{1+\frac{y}{2}}\right) * \text{Macaulay duration}$$

= Approximate percentage change in price
for a 100 basis point change in yield

Dollar duration = (Modified duration) * (Initial price)

Dollar duration is a measure of the dollar price change of the bond for a given yield change. It is also sometimes called the price value of one basis point change.

Portfolio modified duration is approximately the weighted average modified durations of the bonds in the portfolio. It may also be calculated by computing the duration of the entire portfolio's flow of funds, using the internal rate of return that sets the present value of the entire flow equal to the market value of the portfolio.

COMPUTER PROJECTS

1. Use your favorite computer program to compute duration. Compute the duration for the 4% bond, due in 3 years, at a 4% yield, as we did earlier. Now change the yield to 5% and to 6%. What do you expect to happen to the duration? Will it increase or decrease? By how much? A relatively large amount or a relatively small amount? Why?
2. What happens to the duration as yields become very large and very small? Why?
3. Now change Problem 1 to use a 4% bond due in 10 years, but callable at par in 3 years. Price the bond to the par call date when the yield becomes equal to or less than 3%. What happens to the duration as the yield crosses 3%? Why?

TOPICS FOR CLASS DISCUSSION

1. Discuss the various time measures for a flow of funds. Which would you use, and for what purposes would they be useful? What are the advantages and disadvantages of each measure? How well would they measure risk? How well do they conform to the requirements of volatility (or risk) measures?
2. Several people have said that for most individual investors, average life is much easier to compute and to understand than modified duration

and is just as good a measure of risk for their purposes. Do you agree or disagree? What comments do you have on that statement?

PROBLEMS

1. In the two examples in Table 15.1, which cash flow has the longer duration? What do you need to know in order fully to answer this question?
2. Compute the duration for a 5% 4-year bond at a 6% yield.
3. Compute the duration for a 6% 4-year bond at a 5% yield.

Convexity

This chapter presents convexity as a development from duration and as a response to a need for better predictions of price changes as yields change than duration could provide. Convexity meets the needs for this increased prediction precision.

Mathematically, convexity is a natural development and extension of modified duration. We'll cover this factor mathematically in the next chapter, which contains a mathematical development of duration and convexity, as well as the mathematical basis for their use in price change predictions. This chapter presents the equation for convexity, shows how it is used, and gives some examples.

When you finish this chapter, you should understand the equation for convexity, how to compute the convexity of a bond, and how to make the convexity adjustment in estimating a bond price change for a given yield change.

CONVEXITY

The price/yield curve is convex, as illustrated earlier in chapter 6. We can measure this convexity, or degree of curvature, and, as a result, we can use the convexity to improve the predicted price.

The convexity (in periods) is as follows:

$$\text{Convexity} = \frac{1(2)PVCF_1 + 2(3)PVCF_2 + \ldots + n(n+1)PVCF_n}{(1+y)^2 \bullet PVTCF} \qquad \text{Equation 16.1}$$

$$= \frac{\sum_{t=1}^{n} [t(t+1)PVCF_t]}{(1+y)^2 \bullet PVTCF} \qquad \text{Equation 16.2}$$

where

$PVTCF$ = Present value of the total cash flow
$PVCF_t$ = Present value of the cash flow at time t
n = Total number of cash flows
y = Yield at which the cash flows are discounted, per period (Note that in this case y is the yield per period.)

For zero coupon bonds,

$$\text{Convexity} = \frac{n \bullet (n+1)}{(1+y)^2} \qquad \text{Equation 16.3}$$

To calculate convexity in years, we use the following equation:

$$\text{Convexity (in years)} = \frac{\text{Convexity (in periods)}}{k^2} \qquad \text{Equation 16.4}$$

where

k = Number of periods per year (2 in the United States for most calculations)

To compute the approximate percentage price change due to convexity, we use the following equation:

$$\begin{aligned}&\text{Approximate percentage price change due to convexity}\\&= (.5) * (\text{convexity}) * (\text{yield change})^2\end{aligned} \qquad \text{Equation 16.5}$$

Possible convexity's dimension should be time^2, but everybody uses years.

CONVEXITY CALCULATION

Table 16.1 is an example of computing the convexity of a bond. We use the same bond we used in computing the duration: a 4% bond, due in 3 years, at a 4% yield.

Note that in this case, $k = 2$, the number of compoundings per year. Each semiannual period is counted as 1, so the answer was in half years. To change to years, we divided by $k^2 = 4$. If we had measured the periods in half years, going from $t = .5$ to $t = 3$, then the amount in the fourth column in Table 16.1 would have been $t(t + 1/2)$, and the result would have been in years.

Note also that in computing the convexity, we divide by 100 because it is the price of the bond.

ADDITION FOR CONVEXITY

EXAMPLE 16.1. To compute the price change (addition) for a 10 basis point change in yield,

$$(.5)(9.45)(.001)^2 = (.5)(9.45)(.000001) = 0$$

In this case, the convexity change is too small to be used as an adjustment. However, the change exists; it just isn't big enough to be significant in this example. ■

TABLE 16.1 Calculation of Convexity

Period	Cash flow	PVCF	$t(t + 1)$	$t(t + 1)$PVCF
		4% bond due in 3 years at a 4% yield		
1	2.00	1.960784	2	3.921568
2	2.00	1.922338	6	11.534028
3	2.00	1.884644	12	22.615728
4	2.00	1.847690	20	36.953800
5	2.00	1.811462	30	54.343860
6	102.00	90.573080	42	3804.069360
		100.000000		3933.438344

$$\text{Convexity} = \frac{3933.438344}{(1.02)^2 \cdot 100} = \frac{39.33438344}{(1.02)^2} = 37.81$$

(periods)

$$\text{Convexity} = \frac{37.81}{2^2} = 9.45$$

(in years)

EXAMPLE 16.2.   To compute the price change for a 300 basis point change in yield,

$$(.5)(9.45)(.03)^2 = (.5)(9.45)(.0009) = .0042525 = .425\%$$

MODIFIED PRICES FOR CONVEXITY

	Price after duration adjustment	Convexity adjustment	Price after convexity	Actual adjustment	Delta
Decrease yield	108.403+.425		108.828	108.84	.012
Increase yield	91.597+.425		92.022	92.01	−.012

Adding the adjustment for convexity has eliminated 97% of the delta due to using only duration. The adjustment for convexity is positive and changes the symmetry of the adjustment due to duration, bringing the result closer to the actual price.   ■

DOLLAR CONVEXITY

Sometimes people want the dollar price change due to convexity. To compute this, use dollar convexity. The equation is as follows:

$$\text{Dollar convexity} = \text{Convexity} \times (\text{initial price})$$

Then, for the dollar price change due to convexity, we have

$$\text{Dollar price change} = (.5) \times (\text{Dollar convexity}) \times (\text{Yield change})^2$$

Here is an example, from the previous case of our 4% bond, due in 3 years, at a 4% yield.

Previously, we computed the convexity = 9.45, and the price was 100. Then

$$\text{Dollar convexity} = (9.45) \times (100)$$
$$= 945$$
$$\text{Dollar price change} = (.5) \times (945) \times (.03)^2$$
$$= (.5) \times (945) \times (.0009)$$
$$= .42525,$$

in agreement with the calculation shown above, to three decimal places.

MEANING OF CONVEXITY

Convexity is a measure of the curvature of the price/yield curve. Some price/yield curves have relatively little curvature. For example, the price/yield curve of a T-bill maturing in 1 week would have relatively little curvature.

However, the price/yield curve of a long-term, low-coupon bond would have relatively high curvature.

Price/yield curves with relatively low curvature will have low convexity. The 4% bond due in 3 years has relatively low convexity. As a result, the convexity adjustments were relatively small.

A 30-year zero-coupon bond will have relatively high convexity. Convexity adjustments for this bond will be relatively large.

All bonds will have a convexity adjustment of some size. Not all of them will have a large enough adjustment to worry about.

A MISCONCEPTION ABOUT CONVEXITY

Some people think that the convexity adjustment "brings the predicted price up to the price/yield curve." This is wrong. The convexity adjustment brings the predicted price closer to the price/yield curve and therefore is an improvement. It does not bring the predicted price up to the price/yield curve. There will always be an error amount.

PORTFOLIO CONVEXITY

You can use the convexity equation to compute the convexity of a portfolio. But remember that the result will be a statistical measure of the portfolio's characteristics. The portfolio should be composed of bonds similar in the features that are used in computing convexity: coupon and maturity. Otherwise, the measure may not have a real meaning.

CHAPTER SUMMARY

The price/yield curve is generally convex. Convexity is a measure of the degree of convexity (curvature) of this curve.

The equation for convexity (in periods) is:

$$\text{Convexity} = \frac{1(2)PVCF_1 + 2(3)PVCF_2 + \ldots + n(n+1)PVCF_n}{(1+y)^2 \bullet PVTCF}$$

$$= \frac{\sum [t(t+1)PVCF_t]}{(1+y)^2 \bullet PVTCF}$$

The adjustment for years is division by k^2, where k is the number of periods per year (usually 2 in the United States).

The approximate percentage price change due to convexity

$$= (.5) * (\text{Convexity}) * (\text{Yield change})^2$$

Dollar convexity

$$= \text{Convexity} \times (\text{Initial price})$$

COMPUTER PROJECTS

1. Use your favorite math computer program to calculate the convexity of a 90-day T-bill and a 30-year Treasury bond. Print out the shape of the price/yield curves. Compare the two different curves. How do they differ? What is the reason for the difference?
2. Compute the convexity for a 4% 3-year bond at 4%. Then vary the coupon, the maturity date, and the yield. What happens to the convexity? Why?

TOPIC FOR CLASS DISCUSSION

How would you use convexity for portfolio management?

PROBLEM

Compute the convexity for a 4% 3-year bond at 5% and 6% yields, using the example in Table 16.1 as a model. How does the convexity change as the yield changes?

CHAPTER 17

The Mathematical Development of Duration, Convexity, and the Equation to Predict New Bond Prices, Given Yield Changes

This chapter shows the mathematical development of the equations for duration and convexity and uses the Taylor's Series expansion to show the equation for estimation of the predicted bond price, given a change in yield, using duration and convexity. We then discuss the ideas of negative duration and negative convexity and their real-world analogues.

This chapter requires some understanding of elementary calculus. The first year of most calculus sequences covers the required calculus. Most individuals who have studied calculus should have no problem following the development, although it might take a while for the mathematics studied possibly years ago to come back (it took me a while to refresh my knowledge).

When you finish this chapter, you should understand the derivation of duration and convexity, what these terms mean mathematically and in the financial field, and how you might apply them to portfolio management and other situations.

WHY THIS CHAPTER IS IMPORTANT

Many readers, especially those without much formal mathematical training, might wonder why this chapter is even included in the book. Why not simply use the equations given for duration and convexity, without further ado? This section explains why the chapter is included.

Newton and Leibniz independently invented the calculus in the 17th Century, and Taylor developed Taylor's Series in the early 18th Century. If we can connect duration and convexity with this mathematics, we can take advantage of over three hundred years of mathematical development. This does not solely include mathematical knowledge, although that is certainly important. It also includes schools (public and private), colleges and universities, teachers and professors of mathematics, libraries, text and reference books, computer programs, and a whole mathematics infrastructure to spread and make available mathematical knowledge. We can apply a truly large body of mathematical resources to the solution of problems we face.

In addition, stating ideas in mathematical form frequently makes interpretations easy that might otherwise be difficult. You will find, in this chapter, interpretations of negative duration and convexity, and when these situations might occur. These interpretations would be difficult, or even impossible, without the insights available through the use of higher mathematics. In this chapter, the calculus provides us with these insights, which lead us to a better understanding of the subject. But they also lead us to extend our knowledge, and to extend the area of applications of this knowledge.

This is one reason why the study of mathematics is so important.

For many years, I have made these comments part of my two-day fixed income mathematics course. But Wilmott, et al, also refer to this in their excellent book, (page xi), when they comment, in reference to the development of Black-Scholes from the heat diffusion equation, that "we have nearly two centuries of theory on which to call," and "we may consider ourselves to be on well-known territory."

DERIVATION OF DURATION

We recall the equations for duration and modified duration, as presented in Chapter 15:

$$\text{Macaulay duration} = \frac{\sum t \bullet PVCF_t}{k \bullet PVTCF} \qquad \text{Equation 17.1}$$

$$\text{Modified duration} = \frac{Duration}{1+\dfrac{y}{2}}$$

Equation 17.2

Recall also the equation for bond price, given the yield:

$$P = \frac{RV}{\left(1+\dfrac{y}{2}\right)^{N}} + \sum_{i=1}^{N}\frac{\dfrac{c}{2}}{\left(1+\dfrac{y}{2}\right)^{i}} = RV\left(1+\frac{y}{2}\right)^{-N} + \sum_{i=1}^{N}\frac{c}{2}\left(1+\frac{y}{2}\right)^{-i}$$

Equation 17.3

This equation can be differentiated indefinitely with respect to y.

Taking the first derivative of this equation, with respect to y, we have

$$\frac{dP}{dy} = RV(-N)\left(1+\frac{y}{2}\right)^{-N-1}\left(\frac{1}{2}\right) + \sum_{i=1}^{N}\left(\frac{c}{2}\right)(-i)\left(1+\frac{y}{2}\right)^{-i-1}\left(\frac{1}{2}\right)$$

Equation 17.4

Moving the -1, the $(1/2)$, and the factor $[1/(1 + y/2)]$ outside the expression, and multiplying and dividing by P ($P \neq 0$), we have

$$\frac{dP}{dy} = (-1)\left(\frac{1}{2}\right)\left(\frac{1}{1+\dfrac{y}{2}}\right)\left[\frac{N\bullet RV}{\left(1+\dfrac{y}{2}\right)^{N}} + \sum_{i=1}^{N}\left(\frac{c}{2}\right)\frac{1}{\left(1+\dfrac{y}{2}\right)^{i}}(i)\right]\bullet\frac{1}{P}\bullet P$$

Equation 17.5

The expression inside the square brackets, divided by P, is just the Macaulay duration. The $(1/2)$ is the k in the denominator, and the factor $[1/(1 + y/2)]$ converts the Macaulay duration into modified duration. Therefore, we end up with

$$\frac{dP}{dy} = -(\text{Modified duration}) \times P$$

Equation 17.6

Or, expressing modified duration in terms of the other factors,

$$\text{Modified duration} = (-1)\left(\frac{1}{P}\right)\frac{dP}{dy}$$

Equation 17.7

Restating Equation 17.6, we have the value of the first derivative of the price/yield curve as

$$\frac{dP}{dy} = -P(Modified\ duration) \qquad \text{Equation 17.8}$$

NEGATIVE DURATION

When we developed duration as a time measure, we didn't consider the possibility of negative duration. However, we can now look at that concept. Under what conditions could modified duration be negative?

If modified duration is negative, then dP/dY must be positive. The slope of the yield/price curve must be positive. That means that increasing yield must increase price. Under what conditions could that happen?

If the bond has a call feature that is, or soon will be, in effect and is selling at a yield close to the coupon rate, then a decrease in yield will not increase price. The bond is already selling as high as it can sell, given the call. For example, in late June 2001, the United States Treasury had outstanding a $7\frac{7}{8}\%$ bond due in November 2007, but callable at par in November 2002, at a yield (to call date and price) of about 3.47%, while noncallable Treasuries maturing in 2007 were yielding about 4.8%. If yields on 6-year Treasuries (2007 maturity in 2001) fall, the price of this callable bond will not go up much because of the call feature. If the bond were callable in November 2001, at 100, the price would not be much over par, no matter how small the yield was. This bond can have negative duration because rising yields will make it less likely to be called, resulting in a price increase.

Mortgage-backed securities can have similar features. High-coupon mortgage-backed securities have much higher redemptions during periods of low interest rates, as the original borrowers on the mortgages in the high coupon mortgage pools refinance their mortgages. Higher mortgage borrowing rates for these mortgage backed securities will mean somewhat lower redemptions and therefore somewhat higher prices. They could have negative duration for at least part of their price/yield curve.

Options of this sort, which are part of the bond contract, are called embedded options. Bonds with embedded options can have negative duration at some combinations of yield to maturity and time to maturity.

DERIVATION OF CONVEXITY

We recall the equation for convexity (in periods):

$$\text{Convexity} = \frac{1(2)PVCF_1 + 2(3)PVCF_2 + \cdots + n(n+1)PVCF_n}{(1+y)^2 \bullet PVTCF} \qquad \text{Equation 17.9}$$

$$= \frac{\sum [t(t+1)PVCF_t]}{(1+y)^2 \bullet PVTCF}$$

Equation 17.10

where

PVTCF = Present value of the total cash flow
$PVCF_t$ = Present value of the cash flow at time t
n = Total number of cash flows
y = Yield at which the cash flows are discounted, per period (Note that in this case, y is the yield per period).

Remember also that we compute convexity in years as follows:

$$\text{Convexity (in years)} = \frac{Convexity\,(in\,periods)}{k^2}$$

Equation 17.11

where

k = Number of periods per year (2 in the United States for most calculations)

We also recall the equation for the derivative of the price yield curve:

$$\frac{dP}{dY} = (-1)\left(\frac{1}{2}\right)\left[N \bullet RV\left(1+\frac{y}{2}\right)^{-N-1} + \sum_{i=1}^{N} \frac{c}{2}(i)\left(1+\frac{y}{2}\right)^{-i-1} \right]$$

Equation 17.12

Taking the second derivative, we have

$$\frac{d^2P}{dy^2} = (-1)\left(\frac{1}{2}\right)\left[N(-N-1) \bullet RV \bullet \left(1+\frac{y}{2}\right)^{-N-2}\left(\frac{1}{2}\right) + \sum_{i=1}^{N} \frac{c}{2}(i)(-i-1)\left(1+\frac{y}{2}\right)^{-i-2}\left(\frac{1}{2}\right) \right]$$

Equation 17.13

Removing from the square brackets (−1), (1/2), and [1/(1 + y/2)]², and multiplying and dividing by P ($P \neq 0$), we have

$$\frac{d^2P}{dy^2} = \left(\frac{1}{2}\right)^2 \frac{1}{\left(1+\frac{y}{2}\right)^2}\left[\frac{N(N+1)RV}{\left(1+\frac{y}{2}\right)^N} + \sum_{i=1}^{N} \frac{\frac{c}{2}(i)(i+1)}{\left(1+\frac{y}{2}\right)^i} \right] \bullet \frac{1}{P} \bullet P$$

Equation 17.14

Recalling the definition of convexity, we have

$$\frac{d^2P}{dy^2} = P(\text{Convexity})$$

Equation 17.15

NEGATIVE CONVEXITY

Can a bond have negative convexity? This can happen with any differentiable function that has a decreasing first derivative. With bonds, this can happen when the price does not increase much as yield decreases. This can happen with embedded options. These bonds can have negative convexity at certain yields and terms to maturity. The two cases presented earlier, bonds selling up against the call price and high-coupon mortgage-backed securities, can have negative convexity. I suppose you would call this concavity.

TAYLOR'S SERIES EXPANSION

We now construct the equation for developing the bond price change from the Taylor's Series expansion. Recall the Taylor's Series:

$$f(x + \Delta x) = f(x) + f'(x)\Delta x + \frac{f''(x)}{2}(\Delta x)^2 + \cdots \qquad \text{Equation 17.16}$$

Substituting P_1 for the predicted (new) price, P_0 for the original price, and taking derivatives at the point (y_0, P_0), we have

$$P_1 = P_0 + P_0'(\Delta \ yield) + \frac{P_0''}{2}(\Delta \ yield)^2 + \cdots \qquad \text{Equation 17.17}$$

For the percentage value change, we have

$$\frac{P_1 - P_0}{P_0} = P_0'\left(\frac{1}{P_0}\right)(\Delta \ yield) + \frac{P_0''}{2}\left(\frac{1}{P_0}\right)(\Delta \ yield)^2 \qquad \text{Equation 17.18}$$

Substituting, we have

$$\frac{P_1 - P_0}{P_0} = \frac{-P_0(Modified \ duration)(\Delta \ yield)}{P_0} + \frac{\left(\frac{1}{2}\right)P_0(Convexity)(\Delta \ yield)^2}{P_0}$$

$$\text{Equation 17.19}$$

Canceling the P_0, we obtain

$$\frac{P_1 - P_0}{P_0} = -(Modified \ duration)(\Delta \ yield) + \left(\frac{1}{2}\right)(Convexity)(\Delta \ yield)^2$$

$$\text{Equation 17.20}$$

In these latest equations, we disregard the terms beyond the second.

REASONS FOR THE EQUATIONS AND ADDITIONAL FACTORS INTRODUCED IN THE PREVIOUS TWO CHAPTERS

In the previous two chapters, we introduced additional factors without explaining why we used them, why they were there, and what their function was. Here is the explanation, in the light of the calculus derivations and the Taylor's Series expansion.

You can see that bringing in the factor $[1/(1 + y/2)]$ to the Macaulay duration simply converts the duration measure into a close relative of the first derivative of the price/yield curve and makes the equation to predict the new price much easier to understand and to use.

The "k" and "k^2" terms in the denominator of the expressions for duration and convexity are there because we must take the derivative of the term $[1 + (y/2)]$ in the price yield formula. If we have monthly payments, such as for mortgages, instead of semiannual payments, $k = 12$.

The multiplier of $\frac{1}{2}$ in the second term comes from the coefficient in the Taylor's Series expansion, not from the semiannual payments. That term would still be there and would still be $\frac{1}{2}$ if we were analyzing mortgages instead of bonds.

Note that the exponent of the change in the Taylor's Series expansion is 2, so that for positive convexity (the usual case), the adjustment for convexity will always be positive. This adjustment is not symmetric, and it moves the predicted price closer to the actual price on the price/yield curve. However, it doesn't take it all the way. The infinite other terms in the series will do that. But, for financial work, these adjustments are almost always quite small, and nobody worries about them in this case. Other numerical analysis work may require consideration of these terms.

COMPUTER PROJECTS

1. Use your favorite math program to construct a table of sample durations and convexities for a set of bonds. What can you learn from this table about the changes in durations and convexities as the bond parameters change?

2. Use your favorite math program to construct a table of the third term in a Taylor's Series expansion for a set of price/yield calculations for a varied selection of bonds. For example, take 30-year bonds for coupons from 1% to 12% and similar yields. How large is the third term, and what difference does this make?

TOPICS FOR CLASS DISCUSSION

1. Why don't we use more terms in the Taylor's Series expansion for bond price prediction? Why stop at just two terms?
2. Under what conditions might we use additional terms in bond price estimation? What might be the market conditions at those times?

SUGGESTION FOR FURTHER STUDY

Willmott, Paul; Howison, Sam; and Dewynne, Jeff, *The Mathematics of Financial Derivatives*, Cambridge University Press, New York, 1995. For graudate or advanced undergraduate students. Presents modeling of mathematical derivatives from an applied mathematician's point of view. Used as a textbook in several well-known universalities.

CHAPTER 18

Probability and Some Applications to Finance

If you do much work in finance, you should understand at least the rudiments of probability and statistics, and how they can be used in the financial field. Probability has been applied in the actuarial field for many years. However, only relatively recently have the ideas of probability been applied extensively in finance. They can also be applied in project analysis. This chapter and Chapter 20 on variable cash flows show you how you can use these concepts in analysis in these fields.

This is not the place for a complete introduction to probability and statistics. Many fine books exist on these subjects. We present a simple overview of elementary probability concepts, the ideas of discrete and continuous distributions, and some statistical measures, and then we explain how you might use some of these concepts.

ELEMENTARY CONCEPTS IN PROBABILITY: A REVIEW

Probability, in the sense we will use it here, means a measure of the chance of an event, or a set of events, occurring. Probabilities can range from 0 to 1, inclusive. A probability of 0 means that the event is impossible; a probability of 1 means the event is certain. Thus, if p is a probability, then

$$0 \leq p \leq 1$$

EXAMPLES OF PROBABILITIES

Some examples of probabilities familiar to almost everybody. For instance, if you flip a fair coin, there is a 50% chance it will land heads and a 50% chance it will land tails. This is, in fact, the definition of a fair coin. Then, for a fair coin,

$$\text{Probability of heads} = \text{Probability of tails} = \frac{1}{2}$$

If you cast a fair die, you can have one of six outcomes, from one to six inclusive. Each outcome has a $\frac{1}{6}$ probability of occurring.

If you pull one card from a standard, complete deck of cards, you have a $\frac{1}{52}$ probability of pulling any particular card, a probability of $\frac{1}{4}$ of pulling any particular suit, and a probability of $\frac{1}{2}$ of pulling a particular color.

INDEPENDENT EVENTS

Independent events occur when neither event affects the outcome of the other event. Here are some examples.

You toss a coin and record the result, either heads or tails. You throw it again and record the result again. The result of the second toss is not related to the result of the first toss. The two tosses are independent events.

You cast a die and record the result, one of the six outcomes. You cast it again, and again you record the result. The two outcomes are independent.

You pull a card from a complete deck and record the result. You replace the card in the deck and pull again. The two results are independent. However, if you do not replace the first card pulled, the result of the second pull is not independent of the first pull, and the two pulls are not independent events because the results of the first pull affect the results of the second pull.

THE GAMBLER'S FALLACY

Many people think that if a particular outcome does not occur for a long while, the chances of it occurring in the future are increased. For example, they believe that if you toss a fair coin, and it comes up heads ten times in a row, the chances of tossing a tail on the next toss are increased. In the case of independent events, this is untrue and is called "the gambler's fallacy." I have also heard it called "the maturity of chances."

PROBABILITY AS A MATHEMATICAL MODEL

Probabilities represent a mathematical model of the world, similar to the models we discussed in Chapter 3. However, like the other models, they correspond to the world only approximately. For example, a tossed coin could, at least in theory, land on its edge (this was the subject of a *Twilight Zone* episode). Similarly, a die could, at least in theory, split in two when it is cast (this occurred in one of Thomas Hardy's novels). Probably no real coin or die ever manufactured was, or is, exactly "fair." One explanation for the demise of the Long Term Capital Corporation was that the probabilities it used did not correspond closely enough to the real world. This is discussed somewhat more in Chapter 23, on options.

USE OF THE WORD "POPULATION"

You will frequently see the word "population" used in this and other texts. In this context, "population" refers to the set of subjects you are using for probability analysis. This could be people or certain sets of people, but it could also be other collections. For example, when Consumers Union examines the repair rate for automobiles, it examines a population of cars and their repair rates. Each individual car model would be a population, with its own repair rate, and would be a subset of the entire population of automobiles examined in the study.

SOURCES OF PROBABILITIES

Where do we get the probabilities we use, all the time, in various situations? Here are the main sources.

EXPERIENCE

Experience with situations similar to the ones under analysis offers a good source for probabilities. Insurance companies use their experience with insureds to develop premiums. How does an automobile insurance company know that males under age 25 should be charged higher insurance rates for automobile casualty insurance? They have experience with thousands of insureds, of all ages, which shows the need for higher rates and how much the insurance companies must charge. Life insurance companies use their experience to develop mortality, morbidity (sickness), and accident tables, showing the probabilities of death, accidents, and sickness by age. They use the mortality tables to develop life insurance premiums and reserve requirements. Health insurance companies have similar experience for health insurance tables. Workers compensation insurance companies use workplace accident experience to compute premiums.

Insurance companies frequently exchange experience data, or combine it into one large pool of data, for rate-making purposes.

If you use mortality tables, you must use mortality tables that correspond with your requirements, because different classes of people have different mortality rates. You cannot just use a "one size fits all" mortality table. For example, males and females have different mortality rates for the same attained age. Occupations also have different mortality rates; some occupations are particularly dangerous, while others are much safer. Smokers, especially heavy smokers, have higher mortality rates. Persons with a family history of certain diseases, such as cancer, may have higher mortality rates from that disease. Demographic specialists and actuaries study these statistics and develop different mortality tables for different purposes.

The Social Security and some other government agencies in the United States develop life insurance tables for the U.S. population. Life insurance companies could not necessarily use these for insurance purposes because they are taken from a different population than the life insurance insureds.

Other industries have other needs for probabilities. The accounting profession has used experience to develop depreciation schedules. Many businesses use their own projections of asset life to project future replacement needs. For example, the telephone companies have schedules, equivalent to mortality tables, for the expected life of telephone poles in order to project future telephone pole requirements. All of these projections are useful in developing the probabilities of certain events happening. Tables similar to mortality tables can be used for equipment that cannot be repaired when it breaks down, like lightbulbs.

A Look at an Actual Mortality Table

Table 18.1 on page 238 shows the American Experience Mortality Table. This was the first mortality table based on the experience of lives in America.

It was constructed in 1868, based on the experience of the Mutual Life Insurance Company of New York. For many years, almost all life insurance companies in the United States used it to calculate life insurance premiums and reserves. In the 1940s, newer mortality tables, particularly the Commissioners' 1941 Standard Ordinary Table, replaced it. The American Experience Table is now obsolete and no longer accurately measures mortality rates, especially at the younger ages. However, many large companies will still have some business on their books based on the American Experience Table.

The American Experience Table is based on 100,000 alive at age 10, with all dead by age 96.

The table shows the following:

l_x Number living at the start of age x
d_x Number dying at age x
p_x Probability of living through age x
q_x Probability of dying at age x
Note that $P_x + q_x = 1$.

For example, at age 20, 92,637 are alive at the start of the year. The number dying equals $(.007805) \times (92,637) = 723$. The number alive at age 21 equals $92,637 - 723 = 91,914$.

Remember that the selection of the correct mortality tables requires knowledge and careful consideration. You really have to know what you are doing. The American Experience Table is no longer used and has only historical interest. That is one reason we chose it as an example.

Mortality Ratios

A mortality ratio is the relationship between the mortality rate, sometimes due to a specific cause, in a specified population subgroup and the main population. For example, actuaries compute the mortality rates of overweight people and compare them to the mortality rates of the entire population or to the mortality rates of persons with weights within the normal rates. The actuaries would compare mortality rates, by age, for each group. They could then compute the mortality ratio for the two groups, for each age, and for all ages combined.

Such a study might also look at overweight persons and their rates of heart attacks. The study could compare the death rates from heart attacks of overweight persons with heart attack death rates of persons with weights within a normal range, or with the entire population, including the overweight persons. Presumably, overweight persons will have higher heart attack rates, but in any case, the study could show mortality ratios for heart attacks in overweight persons and persons within a normal weight range, or the entire population. Note that the parameters of the study must be carefully defined; overweight and normal weight ranges must be defined, and deaths must be examined to make sure that heart attacks were the only cause. The study also must make sure that allowances are made for the death rates for the actual ages. You cannot compare death rates at high ages for the subgroup with death rates at low ages for the entire population. Death rates for the same ages must be compared. This particular study would measure morbidity ratios, not mortality ratios.

High mortality ratios don't necessarily mean that the death rates are high, but only that the ratios are high. For example, a few years ago, smoking mortality ratios for lung cancer were reported at 500% compared to nonsmokers. This means that the chance of a smoker dying of lung cancer is five times the chance of a nonsmoker dying of lung cancer. But lung cancer death rates in the population as a whole are relatively low, and quite low at young ages. Even a 500% mortality ratio won't result in a very high absolute mortality rate, but only in a comparatively high mortality rate.

The surgeon general's 1990 report on "The Health Benefits of Smoking Cessation" shows a large number of mortality (and other) ratios. Here are a few. All of these are taken from the report; the American Cancer Society is the source for many of them.[1]

One table (page 78) shows mortality ratios, broken down by sex and daily cigarette consumption, for current smokers and former smokers, broken down by duration of abstinence at the time of the study. Male current smokers of 1 to 20 cigarettes per day show a mortality ratio of 2.22. In this case, the standard population is males who never smoked and the test population is males who now smoke from 1 to 20 cigarettes daily. For female former smokers, with an abstinence duration of 16 years or more and a former consumption of 1 to 19 cigarettes daily, the mortality ratio is 1.01. In this case, the standard population is females who have never smoked, and the test population is females who have not smoked for 16 or more years and smoked 1 to 19 cigarettes daily when they did smoke.

The same surgeon general's report shows mortality ratios for specific diseases, such as lung cancer. For example, current male smokers of 1 to 20 cigarettes daily show a mortality ratio from lung cancer of 18.8 relative to males who never smoked. In this case, the standard population is males who never smoked, the test population is males who now smoke 1 to 20 cigarettes daily,

[1]These statistics are all taken from the Surgeon General's Report, and are based on unpublished tabulations from the American Cancer Society. See the chapter's end for the citation.

and the measure is of deaths from lung cancer only, not all deaths. Females who formerly smoked 1 to 20 cigarettes daily but have abstained from smoking for 16 years or more show a lung cancer mortality ratio of 1.4. The standard population is females who never smoked, the test population is females who smoked 1 to 20 cigarettes daily but have abstained for 16 or more years, and the measure is of lung cancer deaths only.

You can also apply the mortality ratio concept to other fields. For example, Consumers Union computes repair rates for various models of cars, based on their reports, and compares them to other makes and models of cars. Businesses could look at repair and maintenance costs for their various makes of machines they use and make decisions on which brands to buy.

THEORETICAL CONSIDERATIONS

Theoretical considerations can develop a considerable body of probabilities. We looked at some of these earlier in this chapter when we discussed the probabilities of various outcomes for coin tosses, dice casting, and card pulling. Further analysis can develop the probability of passing in a crap game or of losing at blackjack, and can compute the correct odds at roulette. Casinos use these probabilities all the time. Probability study got a start in the 17th century when some gamblers approached the French mathematician and philosopher Blaise Pascal with a gambling problem, although some work on probability had been done earlier.

Other theoretical sources could come from mathematical models similar to those discussed in Chapter 3. For example, the life insurance industry has developed some mathematical models of human mortality and has had some success in modeling mortality rates.

Mathematical functions that generate probabilities can also be used. These are called probability distribution functions (sometimes they are also known as probability-generating functions). We will look at several of these functions later in this chapter.

SIMULATION ANALYSIS

Sometimes you can develop a set of probabilities by developing a model of an operation or activity of some sort. You enter numerical inputs into the model using a table of random numbers or a computer random number-generating program to determine numerical outputs. You can then determine the probabilities of the various outputs. Many statistics books, and most books of statistical tables, contain tables of random numbers. Computer programs that generate random numbers also exist, although whether the numbers generated by a computer can ever be truly random remains uncertain, at least to me.

JUDGMENT

Often, experienced people can call on their own experience to give probabilities of various events. These probability suggestions can often be quite good.

PROBABILITY DISTRIBUTION FUNCTIONS

Many probability distributions can be expressed as a mathematical equation. This allows mathematical modeling of the probabilities and allows mathematical prediction of the outcomes of a number of trials. Mathematical functions that do this are called "probability distribution functions." Here is an example.

THE BINOMIAL DISTRIBUTION

Suppose you have a trial that can have just two outcomes, called "success" and "failure." After n trials, how many successes and how many failures will you have? This is called the binomial distribution.

You have just two outcomes, success, denoted by p, and failure, denoted by q. Then $p + q = 1$, and the equation for the outcomes, after n trials, is

$$(p+q)^n = p^n + np^{n-1}q + \left(\frac{n(n-1)}{2}\right)p^{n-2}q^2 + \ldots + q^n = \sum_{i=0}^{n}\left(\frac{n!}{i!(n-i)!}p^{n-i}q^i\right)$$

where

$$n - i = \text{Number of successes}$$
$$i \quad = \text{Number of failures}$$
$$n! \quad = n(n-1)(n-2)\ldots(3)(2)(1) \text{ and } 0! = 1$$

EXAMPLE 18.1.   You flip a fair coin three times. What are the probabilities for different outcomes of heads (H) and tails (T)?

You have four different outcomes, HHH, HHT, HTT, and TTT. The order of the outcomes doesn't matter, so that HHT, HTH, and THH are considered the same: two heads and one tail.

Here is the general equation for the outcome:

$$(p+q)^3 = (p^3 + 3(p^2)(q) + 3(p)(q^2) + q^3)$$

In the case of a fair coin, p = q = .5, and we have

$$(.5+.5)^3 = \left(.5^3 + 3(.5^2)(.5) + 3(.5)(.5)^2 + .5^3\right) = .125 + .375 + .375 + .125$$

The probabilities for each outcome, in tabular form, are

HHH	.125
HHT	.375
HTT	.375
TTT	.125
Total	1.000

You can also develop these probabilities in a somewhat different way. To toss three heads (HHH), you must toss a head, with a probability of .5, three times. The probability of doing this is $(\frac{1}{2})(\frac{1}{2})(\frac{1}{2}) = \frac{1}{8}$.

To achieve a HHT, you must throw heads twice and tails once. Once more, the probability of doing this is $(\frac{1}{2})(\frac{1}{2})(\frac{1}{2}) = \frac{1}{8} = .125$. However, there are three ways of throwing two heads and one tail: HHT, HTH, and THH. Each of these has a $\frac{1}{8}$ probability, so the chance of throwing two heads and one tail $= \frac{1}{8} + \frac{1}{8} + \frac{1}{8} = \frac{3}{8} = .375$.

Calculations for one head and two tails, and for three tails, are similar. ∎

EXAMPLE 18.2.   You flip an unfair coin three times. The coin has a probability of heads 60% of the time (p = .6) and tails 40% of the time (q = .4). What are the probabilities of the outcomes? Here is the equation, with p = .6 and q = .4:

$$(.6+.4)^3 = (.6^3 + 3(.6^2)(.4) + 3(.6)(.4^2) + .4^3) = .216 + .432 + .288 + .064$$

The probabilities for each outcome, in tabular form, are

HHH	.216
HHT	.432
HTT	.288
TTT	.064
Total	1.000

Once more, the order doesn't matter, only the total of heads and tails respectively. ∎

CONTINUOUS PROBABILITY DISTRIBUTIONS

Continuous probability distributions also exist. In these cases, the probability distribution function defines the probability of the event occurring within an interval as the area under the function's curve in that interval divided by the total area under the curve; the function can usually be changed so that the area under the entire curve equals 1. The event is defined as an interval (a_0, b_0) on the x-axis, and the probability is the area under the curve on the interval (a_0, b_0).

A probability distribution function for a continuous function is a function $f(x)$ where

1. $f(x) \geq 0$
2. $\int_{-\infty}^{\infty} f(x)dx = 1$
3. $\int_{a}^{b} f(x)dx = Prob(a \leq x \leq b)$

You can see that the probability of x being in the interval $a \leq x \leq b$ is the value of the area under the curve in the interval $a \leq x \leq b$.

Here are three examples of continuous probability functions.

EXAMPLE 18.3.   Figure 18.1 shows a hypothetical probability distribution on the interval [a,b]. All events occur within the interval [a,b], inclusive. The area under the curve, from a to b inclusive, is assumed to equal 1. ■

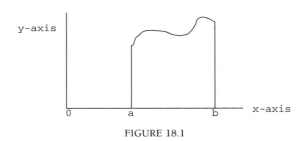

FIGURE 18.1

EXAMPLE 18.4.   This illustration shows the uniform distribution. The interval goes from (0,0) to (1,0) on the x-axis at a height of 1. The chance of events in the interval [a,b] occurring is the area of the area with diagonal lines under the curve. In this case, the probability is (b − a).   ■

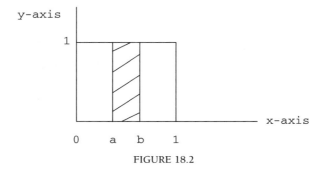

FIGURE 18.2

EXAMPLE 18.5.   Figure 18.3 shows the graph for a probability density function shaped as a triangle from the origin to the point (1,2). The probability density function will be

$$y = 2x$$

You can see that the total area under the curve

$$= \left(\frac{1}{2}\right)(\text{base})(\text{height})$$
$$= \left(\frac{1}{2}\right)(1)(2)$$
$$= 1$$

The probability of occurrence of an event located in the interval (a,b) is shown by the area with diagonal lines in it. The chance of the event occurring is

$$= \left(\frac{1}{2}\right)(b)(2b) - \left(\frac{1}{2}\right)(a)(2a)$$
$$= b^2 - a^2 \qquad \blacksquare$$

THE NORMAL DISTRIBUTION

The most important continuous distribution, for most people, is the normal distribution, also known as the familiar "bell-shaped curve." The equation for the normal distribution is

$$f(x) = \frac{1}{c} e^{-\frac{1}{2}\left(\frac{x-a}{b}\right)^2}$$

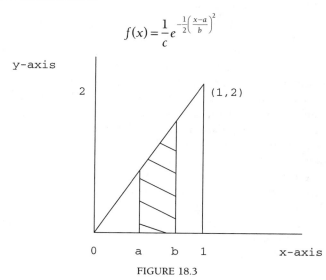

FIGURE 18.3

where a, b, and c are parameters that make f(x) a probability frequency function. c must be chosen to make the area under the curve equal to 1.

The normal distribution equation can also be stated as

$$f(x) = \frac{1}{\sigma\sqrt{2\pi}} e^{-\frac{1}{2}\left(\frac{x-\mu}{\sigma}\right)^2}$$

where μ is the mean, and σ is the standard deviation of the function. Mean and standard deviation are defined later in this chapter.

The normal distribution is well known, widely studied, widely taught, and has attractive mathematical features. Mathematicians like to teach it. This means that when you study bell-shaped curves, you need to take special care.

Not all "bell-shaped" curves are normal distributions. Whenever you work with bell-shaped curves, you should make sure that the normal distribution equation accurately describes them, especially in the "tails," where the probabilities of occurrence are small but the individual effects can be quite large.

STATISTICS AND STATISTICAL ANALYSIS

Statistics applies probability concepts to a population to obtain information about the population. In some uses, the purpose is to obtain information about the population for planning purposes. For example, if you are planning a new product for homeowners, you need information about homeowners who might use the product and the homes in which that the product might be used. You might have an interest in home size, owner's age, number of children in the house, family income, and similar information. This information would be presented in statistical form, such as averages and variations from the average.

As another example, an analyst might calculate the mean height of a population, or the mean weight, and use this information to design clothing, furniture, automobiles, or other products that the population might use. The analyst might look at dispersions about the mean to find out the extent of the market for different sizes of the product, or how many different sizes might be required to meet the population's expected demands.

Another common use is decision making. For example, you run a credit card operation and Pat Jones has applied for a card. You must decide whether or not to issue Pat a credit card, and, if you do issue a card, you must determine what credit limits you will impose. Credit card companies make these decisions all the time, both on new applicants and on present cardholders, either to increase or decrease credit limits, to institute or change annual charges, or even to cancel the credit card.

The two most commonly used pieces of information about a population are measures of central tendency and dispersion. Usually, these are the mean and the standard deviation, or variance. We'll discuss these later on. Decision making, of the sort applied to the credit card applicants discussed earlier, is done by testing and setting confidence limits. These concepts are outside the scope of this book. Almost any introductory statistics book will cover them.

MEASURES OF CENTRAL TENDENCY

Three main measures of central tendency exist: the mean, the median, and the mode. All have some uses, but mathematical statistics almost always uses the mean. The mean corresponds to the commonly used average and is the average for many populations. The median is the observation (or event) at the midpoint. Half the observations are greater, and half the observations are smaller than the median. The mode is the most common occurrence in the set of observations.

EXAMPLE 18.6.   Here is a population of nine observations:

$$4, 4, 4, 5, 6, 8, 10, 16, 24$$

These observations could possibly represent incomes in a firm with nine employees, or ages of children in a family, including cousins. In this case,

The mean = (4 + 4 + 4 + 5 + 6 + 8 + 10 + 16 + 24)/9 = 9
The median = 6(4 observations under 6, 4 over 6)
The mode = 4 (the most common observation)

You can see that the three measures can have quite different values for the same population.

Sometimes data can be grouped, or accumulated. In Example 18.6, the population has three observations of four. These could be grouped together in the following way to form a weighted average:

The mean = ([3*4] + 5 + 6 + 8 + 10 + 16 + 24)/9 = 9

Here is the equation for the mean:

$$\bar{x} = \frac{1}{n} \sum_{i=1}^{k} x_i f_i$$

where

n = Total number of observations
k = Number of different categories
x_i = Number of observations in category i
f_i = Size of occurrences in category i

In Example 18.6,

n = 9 (observations)
k = 7 (different categories)
x_i = 3 for the first category (the three 4s) and 1 for all the other categories
f_i = 4, 5, 6, 8, 10, 16, 24 (the original observations, without repetitions)

The mean is also called the first moment about the origin and often the expected value. You can see that in the case of a set of occurrences, which are all different, the mean is the familiar average. ■

When you use the mean, as in all use of statistical measures, you cannot just apply the measures without some thought. Here is an example:

EXAMPLE 18.7.   A population of 10 persons has the following annual incomes:

9 persons have a $10,000 annual income
1 person has a $1,000,000 annual income

The mean = ([9 ∗ $10,000] + 1 ∗ $1,000,000)/10 = $1,090,000/10 = $109,000.

Using the mean in this case is misleading; nobody comes close to the mean income.

This might actually represent the income distribution in some countries. Economists have developed their own measures of income distribution.

In some countries, such as Brazil or India, hundreds of mutually unintelligible languages are spoken. Simply applying the mean might result in a small number of speakers for each language. Actually, in Brazil, most people speak Portugese, although reportedly hundreds of Indian languages are spoken in the Amazon River basin. In India, a large minority speak Hindi, although reportedly hundreds of other languages are spoken as well. ■

MEASURE OF DISPERSIONS

Two main measures of dispersion exist: the standard deviation and the variance. The standard deviation equals the square root of the variance. Here is the equation for the variance, denoted by s^2:

$$Variance = s^2 = \frac{1}{n} \sum_{i=1}^{k} (x_i - \bar{x})^2 f_i$$

where $\bar{x}$ = the mean and the other symbols have the same meaning as in the equation for the mean. The standard deviation = s = the square root of the variance. The variance is also called the second moment about the mean.

APPLICATIONS TO INSURANCE

Millions of Americans buy insurance for various purposes, including automobile liability insurance, house insurance, personal liability insurance, medical insurance, life insurance, and other purposes. Yet many of these people may not have a good understanding of insurance and how it works. This section is an effort to overcome this lack.

Suppose you own a home worth $200,000. If your home burns down, you have suffered a $200,000 loss. You cannot afford that, so you buy fire (and other) insurance on your home. If your home burns down, the insurance company pays you for the loss.

Suppose there is a $\frac{1}{1000}$ chance that your home will burn down in the next year. The insurance company might charge you $(\frac{1}{1000})(\$200,000) = \$200$, plus perhaps another $200 for loading, to insure you against this loss. You pay the $400. If the house burns down, you have suffered no economic loss (except, of course, the expense of the insurance premium).

This example illustrates the main points and purposes of insurance. The loss event being insured against (the fire in this case) should be large ($200,000 in this case). The chance of the event happening should be small ($\frac{1}{1000}$ in this case). The cost of the insurance should be "moderate," $400 in this case. The net cost, in this case, is the expected value, $200. The insurance company added another $200 to cover various costs of running the business. This is the kind of event that insurance is designed for. In slightly different words, you insure against things that hardly ever happen, but are very expensive when they do happen.

The insurance covers only the economic value of the item insured. You might have a ring that belonged to your great-grandmother and was given to you by your grandmother on her deathbed. The ring has a very great sentimental value to you. The cash value of the ring is $10. You can insure the ring for $10. The emotional value cannot be insured. If the ring is lost or stolen, you can receive the $10, but not anything for the emotional value you have lost.

Insurance is not meant to cover events that have a high probability of occurring or events that have a cost to insure that is more than "moderate." Here are some examples.

EXAMPLE 18.8.   Each week you spend about $75 on food. This is a planned expense, and you pay it out of your income. Insurance is not meant for planned, continuing expenses like this.   ■

EXAMPLE 18.9.   *Collision insurance deductible.* Many people have collision insurance on their car. If they get into an accident, the insurance will pay regardless of who is at fault. But many accidents involve relatively small damage. As a result, many collision insurance policies have a "deductible"; accidental amounts less than or equal to the deductible amount are not paid, and the insurance company reduces the claims payment by the amount of the deductible for claims that exceed the deductible. This means that small claims, which are likely to be more frequent, will not be paid. This is in line with the fundamental principles of insurance. The small, frequent claims are not insured against. Presumably, the insured makes these payments out of regular income. ▪

EXAMPLE 18.10.   *Life insurance at an advanced age.* When people age, the chance of dying in the next year increases. At advanced ages, the chance can become quite high. As prospective death rates increase, insurance premiums increase, and the need for insurance decreases. Instead, people worry more about estate planning. ▪

EXAMPLE 18.11.   *Medical insurance during a terminal illness.* One sometimes hears the phrase, "He is dying, so he really needs insurance." Insurance doesn't apply to cases like this. The person needs cash. The requirements, while possibly large, have a very high probability, almost a certainty, of occurring. ▪

EXAMPLE 18.12.   You want to cross a rickety bridge and wish to insure against loss. The possible loss is $1 million because you are carrying valuable jewelry. The insurance company figures that the chance of bridge collapse is .1, and it charges you $100,000 plus loading for the insurance. You think the charge is too high, it is more than "moderate," and you don't buy the insurance. You go around by another way. ▪

CHAPTER SUMMARY

Probability is the chance p of a given event occurring.

$$0 \le p \le 1$$

$p = 0$ means the event's occurrence is impossible.
$p = 1$ means the event's occurrence is certain.
Independent events occur when neither event affects the other.
The gambler's fallacy states that if a particular outcome has not occurred for a long period of trials, it is more likely to appear in the next trial.
Probability can be a mathematical model of the real world.
Sources of probabilities include experience, theoretical considerations, simulation analysis, and judgment.
Probability distribution functions express probabilities as a mathematical equation.

The binomial distribution expresses the results of repeated trials, each of
which has only two possible outcomes.

The equation for the binomial distribution is

$$(p+q)^n = p^n + np^{n-1}q + \left(\frac{n(n-1)}{2}\right)p^{n-2}q^2 + \ldots + q^n = \sum_{i=0}^{n}\left(\frac{n!}{i!(n-i)!}p^{n-i}q^i\right)$$

where

$$\begin{aligned}
n-i &= \text{the number of successes} \\
i &= \text{the number of failures} \\
n! &= n(n-1)(n-2)\ldots(3)(2)(1) \text{ and } 0!=1
\end{aligned}$$

The equations defining continuous probability distribution functions are

1. $f(x) \geq 0$
2. $\int_{-\infty}^{\infty} f(x)dx = 1$
3. $\int_{a}^{b} f(x)dx = Prob(a \leq x \leq b)$

Probability for a continuous distribution for an event occurring in the
interval (a_0, b_0) is the area under the distribution curve on the interval
(a_0, b_0).

The most important continuous distribution is the normal distribution.

The equation for the normal distribution is

$$f(x) = \frac{1}{\sigma\sqrt{2\pi}}e^{-\frac{1}{2}\left(\frac{x-\mu}{\sigma}\right)^2}$$

where μ is the mean, and σ is the standard deviation of the function.

Statistics applies probability concepts to a population.

The purposes are to obtain information about the population or to make
decisions based on this information.

Two main information items about a population are measures of central
tendency and dispersion.

The main measures of central tendency are mean, median, and mode.

The equation for the mean is

$$\bar{x} = \frac{1}{n}\sum_{i=1}^{k}x_i f_i$$

where

$$\begin{aligned}
n &= \text{Total number of observations} \\
k &= \text{Number of different categories} \\
x_i &= \text{Number of observations in category i} \\
f_i &= \text{Size of observations in category i}
\end{aligned}$$

The measures of dispersion are variance and standard deviation.
The equation for the variance is

$$Variance = s^2 = \frac{1}{n} \sum_{i=1}^{k} (x_i - \bar{x})^2 f_i$$

where $\bar{x}$ = the mean, and the other symbols have the same meaning as in
the equation for the mean.
The standard deviation = s = the square root of the variance.
Insurance applies when the loss is large, the chance of the loss occurring
is small, and the expected loss, or premium, is moderate.
Insurance covers only economic loss.

COMPUTER PROJECTS

1. Use your favorite computer mathematics program to display a normal
distribution. Then change the mean and/or the standard deviation, but
keep the distribution normal. What must happen to the distribution
for the "normality" to remain?
2. Use your favorite computer mathematics program to develop a curve
that is generally "bell-shaped," but is not a normal distribution. What
characteristics does it have? Can the area under a bell-shaped curve be
infinite? If so, how would you know it in actual distributions of data?

TOPICS FOR CLASS DISCUSSION

1. Look at the observations in Problem 2. What conclusions might you
draw about the population just from an inspection of the data? How
would you go about doing a statistical analysis of the population for
the purpose of gathering information about the population? What
examples of this type of collection can you think of from your own
experience?
2. Can you suggest other measures of central tendency or dispersion?
What measures might you suggest? How would you compute these
measures?

PROBLEMS

1. You flip an unfair coin four times. The chance of a head is .55, and
the chance of a tail is .45. What is the probability of throwing two
heads and two tails?

2. Here is a population of observations. What are the mean, the median, the mode, the variance, and the standard deviation?

3, 4, 4, 5, 5, 5, 6, 7, 9, 10, 10, 11, 12, 15, 18, 18, 18, 18, 26, 30, 32

SUGGESTIONS FOR FURTHER READING AND STUDY

Many fine books exist on probability and statistics. Here are three old-timers that are still in print and that I have found useful. In various editions, they have been around for decades; they must be doing something right.

Feller, William, *Introduction to Probability Theory and Its Applications*, Volume 1, 3rd ed., Wiley, New York, 1968; Volume 2, 2nd ed., Wiley, New York, 1991.

Hoel, Paul G., *Introduction to Mathematical Statistics*, 5th ed., Wiley, New York, 1984.

Hogg, Robert V., and Craig, Allen T., *Introduction to Mathematical Statistics*, 5th ed., Prentice-Hall, Englewood Cliffs, NJ, 1994.

The statistics on smoking mortality ratios are taken from: U.S. Department of Health and Human Services. *The Health Benefits of Smoking Cessation*. U.S. Department of Health and Human Services, Public Health Service, Centers for Disease Control, Center for Chronic Disease. Prevention and Health Promotion, Office on Smoking and Health. DHHS Publication No. (CDC) 90-8416, 1990.

For more information on the mathematics and statistics of this area, also see the following, especially Chapter 3: U.S. Department of Health and Human Services. *Reducing the Health Consequences of Smoking, 25 years of Progress, A Report of the Surgeon General*. U.S. Department of Health and Human Services, Public Health Service, Centers for Disease Control, Center for Chronic Disease Prevention and Health Promotion, Office on Smoking and Health, DMS Publication No. (CDC) 89-8411, 1989. Both of these publications are available in many public libraries.

TABLE 18.1

AMERICAN EXPERIENCE MORTALITY TABLE

Based on 100,000 living at age 10, giving: l_x, number of living; d_x, number of deaths; p_x, probability of living; q_x, probability of dying for age x from 10 to 95.

x	l_x	d_x	p_x	q_x	x	l_x	d_x	p_x	q_x
10	100000	749	.992510	.007490	55	64563	1199	.981429	.018571
11	99251	746	.992484	.007516	56	63364	1260	.980115	.019885
12	98505	743	.992457	.007543	57	62104	1325	.978665	.021335
13	97762	740	.992431	.007569	58	60779	1394	.977064	.022936
14	97022	737	.992404	.007596	59	59385	1468	.975280	.024720
15	96285	735	.992366	.007634	60	57917	1546	.973307	.026693
16	95550	732	.992339	.007661	61	56371	1628	.971120	.028880
17	94818	729	.992312	.007688	62	54743	1713	.968708	.031292
18	94089	727	.992273	.007727	63	53030	1800	.966057	.033943
19	93362	725	.992235	.007765	64	51230	1889	.963127	.036873
20	92637	723	.992195	.007805	65	49341	1980	.959871	.040129
21	91914	722	.992145	.007855	66	47361	2070	.956293	.043707
22	91192	721	.992094	.007906	67	45291	2158	.952353	.047647
23	90471	720	.992042	.007958	68	43133	2243	.947998	.052002
24	89751	719	.991989	.008011	69	40890	2321	.943238	.056762
25	89032	718	.991935	.008065	70	38569	2391	.938007	.061993
26	88314	718	.991870	.008130	71	36178	2448	.932335	.067665
27	87596	718	.991803	.008197	72	33730	2487	.926267	.073733
28	86878	718	.991736	.008264	73	31243	2505	.919822	.080178
29	86160	719	.991655	.008345	74	28738	2501	.912972	.087028
30	85441	720	.991573	.008427	75	26237	2476	.905629	.094371
31	84721	721	.991490	.008510	76	23761	2431	.897689	.102311
32	84000	723	.991393	.008607	77	21330	2369	.888936	.111064
33	83277	726	.991282	.008718	78	18961	2291	.879173	.120827
34	82551	729	.991169	.008831	79	16670	2196	.868266	.131734
35	81822	732	.991054	.008946	80	14474	2091	.855534	.144466
36	81090	737	.990911	.009089	81	12383	1964	.841395	.158605
37	80353	742	.990766	.009234	82	10419	1816	.825703	.174297
38	79611	749	.990592	.009408	83	8603	1648	.808439	.191561
39	78862	756	.990414	.009586	84	6955	1470	.788641	.211359
40	78106	765	.990206	.009794	85	5485	1292	.764448	.235552
41	77341	774	.989992	.010008	86	4193	1114	.734319	.265681
42	76567	785	.989748	.010252	87	3079	933	.696980	.303020
43	75782	797	.989483	.010517	88	2146	744	.653308	.346692
44	74985	812	.989171	.010829	89	1402	555	.604137	.395863
45	74173	828	.988837	.011163	90	847	385	.545455	.454545
46	73345	848	.988438	.011562	91	462	246	.467532	.532468
47	72497	870	.988000	.012000	92	216	137	.365741	.634259
48	71627	896	.987491	.012509	93	79	58	.265823	.734177
49	70731	927	.986894	.013106	94	21	18	.142857	.857143
50	69804	962	.986219	.013781	95	3	3	.000000	1.000000
51	68842	1001	.985459	.014541					
52	67841	1044	.984611	.015389					
53	66797	1091	.983667	.016333					
54	65706	1143	.982604	.017396					

Reprinted with permission. From *Mathematical Tables from Handbook of Chemistry & Physics, Tenth Edition.* Copyright CRC Press, Boca Raton, Florida.

CHAPTER 19

The Term Structure of Interest Rates, the Expectations Hypothesis, and Implied Forward Rates

With this chapter, we begin to look at a bond in a different way. Previously, starting in Chapter 6, we looked at it as a series of interest payments, coupled with a final maturity payment. We showed how to compute the price, given the yield.

Many people, and probably most individual investors, look at bonds that way. A bond yields cash income, which the owner spends, and the bond returns the invested principal at some future time. The owner then reinvests the maturing principal.

However, some people, and possibly most managers of large bond portfolios, view a bond as a series of zero-coupon instruments, with each payment of interest and principal as a separate cash flow. Each of these cash flows could be evaluated independently.

We suggested this approach a little bit in Chapter 6 when we stated that the owner of a premium bond could consider each amortized principal amount as a separate investment, maturing at the time of the interest payment, and also suggested that, in times of a normal, ascending yield curve, the owner would have made a short-term investment at a higher, long-term rate. This would generally be advantageous, although in the case of most premium bonds the amounts are relatively small.

In the chapter on present values, we also looked briefly at a set of cash flows, which were not necessarily equal.

This chapter continues that approach. It shows first how you can use one of the term structure hypotheses to compute a set of yields to evaluate a set of cash flows. It then shows how to adjust these yields to produce a set of yields suitable for your analysis purposes. It develops the equation for implied forward rates, once more using the expectations hypothesis. It then presents other term structure hypotheses, in order to make sure first, that you understand that other hypotheses exist, and, second, that you have some idea of the hypotheses themselves and what they mean. One of these other hypotheses, the risk premium hypothesis, is related to the use of duration, which we discussed in Chapter 15.

When you finish this chapter, you should understand how to bootstrap a spot rate curve for Treasuries and how to compute spot yields for different types of bonds in order to evaluate these bonds as well as for your further analytical purposes.

THE TERM STRUCTURE OF INTEREST RATES

If you look at even a relatively complete market report for United States Treasuries, you will observe that securities with different maturities sell at different yields. For example, in early December of 2001, 6-month Treasuries were yielding about 1.78%; 1-year Treasuries were yielding about 2.02%, with yields generally increasing out to about 30 years. Table 19.1 shows the yields for selected maturities.

You can see that, in general, as time to maturity increases, the yield to maturity also increases.

The relationship between time to maturity and yield to maturity is called the term structure of interest rates. Economists have studied the term structure of interest rates for more than 100 years now and have proposed several hypotheses to explain it. We will look at one of these hypotheses, called the expectations hypothesis.

Remember that all term structure hypotheses claim to describe some aspects of human behavior in the securities markets.

SHAPES OF YIELD CURVES

Four main types of yield curves exist: the normal, or ascending, yield curve; the inverted yield curve; the flat yield curve; and the hump-backed yield curve.

TABLE 19.1 Treasury Yields by Maturity,
December, 2001

Time to maturity	Approximate yield
6 mos.	1.78%
1 year	2.03
2 years	2.77
5 years	3.96
10 years	4.65
20 years	5.41
30 years	5.21

Yield

Time to maturity

FIGURE 19.1 Ascending, or normal, yield curve. Yields increase as time increases, although sometimes very long-term yields may decrease a little.

The normal, or ascending, yield curve occurs most of the time, which is the reason it is called "normal" (see Figure 19.1).

The inverted yield curve shows yields decreasing as time to maturity increases. Figure 19.2 displays an inverted yield curve.

The flat yield curve has no change in yield as time to maturity increases. Figure 19.3 displays a flat yield curve.

The hump-backed yield curve rises to a point and then declines again. Usually the high point is in the 7- to 8-year maturity range, at least for Treasuries. Figure 19.4 displays a hump-backed yield curve.

THE EXPECTATIONS HYPOTHESIS

The expectations hypothesis states that the yield curve contains within itself its own prediction of future interest rates. It states that investors are indifferent to the time to maturity. With this assumption, we can compute these forecasted rates using the yields from the bonds trading in the market, and we can

FIGURE 19.2   Inverted yield curve.

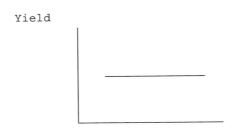

FIGURE 19.3   Flat yield curve.

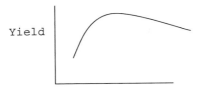

FIGURE 19.4   Hump-backed yield curve.

also compute the implied rates for zero-coupon bonds. Here is a simple example for teaching purposes.

Suppose we have the following term structure for bonds of 1- and 2-year maturities:

Time	Yield
1 year	3.00%
2 years	4.00%

Suppose an investor is willing to make a 2-year investment. He can either buy a 2-year security and hold it to maturity or buy a 1-year security and, when it matures in 1 year, buy another 1-year security. The expectations hypothesis states that the investor is indifferent between the two investment plans. That means that the two plans are of equal present value to him.

Let's analyze and compare the two alternatives. In doing this, for teaching purposes we will not use compound interest equations at first.

The investor could choose to buy the 2-year security. If he does, he will receive 4 each year on an investment of 100, or a total return of 8 over the 2 years.

If he chooses the first option, he will receive three during the first year and an unknown amount, x, during the second year. The expectations hypothesis says that the sum of the two amounts will equal the total earned on the 2-year security. In other words, a table showing the earnings, by year, will be as follows:

Year	Amount earned	
	Choice 1	Choice 2
1	3	4
2	x	4
Total	3 + x	8

Then $x = 8 - 3 = 5$. That is, the expectations hypothesis says that the market expects that 1 year from now, the 1-year rate will be 5%. Assuming the expectations hypothesis, we can compute the expected rate. This is called the 1-year forward rate, 1 year from now.

IMPLIED SPOT RATES AND BOOTSTRAPPING A SPOT YIELD CURVE

In Chapter 6, we developed the equation for a bond price, given the yield. We applied the same yield to all the payments to develop the present value (price plus accrued interest) of the bond. This gave us one equation in one unknown, and we could solve the equation.

But suppose we want to discount each of the payments at a different yield. In the case of a 30-year bond, we would have 60 or 61 payments, depending on whether we discounted the last interest payment and the maturity amount at the same or at different rates. We would have, as a result, 60 or 61 possibly different yields. These individual yields are called the "spot rates" and apply to each individual bond interest or maturity payment, each considered as an independent, zero-coupon security, maturing on the payment date. If the yields apply to Treasuries, they are called a "Treasury spot rate curve" or the "Treasury spot rates." How would we compute these rates?

COMPUTING THE SPOT RATES

We use a method called "bootstrapping" the spot yields. Bootstrapping works in the following way.

Suppose you have a 10-year Treasury bond, and you want to compute the spot rates. You will price both the 10th year interest payment and the final maturity amount at the same yield. How would you compute this yield? You can't right now, because each of the 20 payments would have its own individual yield. You therefore have one equation with 20 unknowns and an infinite number of solutions.

But suppose you already know the first 19 yields, never mind how for the moment. You then have one equation, with just one unknown, the 20th yield, and you can compute this yield. We therefore start with one or two yields we can take from the marketplace and then construct, maturity by maturity, the spot rate curve.

Suppose we know the 6-month rate and the 1-year rate for zero-coupon Treasuries. We then take a Treasury maturing in $1\frac{1}{2}$ years. We have a total of three payments, a present market price of the security, and rates to evaluate the present value of the first and second payments. We can evaluate the present value of the first two (coupon) payments using the two spot rates we already have and we know the market price for the security, so we have one equation in one unknown, which is the spot yield for the third payment (coupon and maturity combined). We solve for the third yield, using a numerical analysis technique.

We continue with a 2-year maturity and compute, in similar fashion, the 2-year yield. We can continue on with this process as long as we have semiannual maturities. If we have only annual maturities, we can still bootstrap an annual series of spot rates.

Note that you could start bootstrapping with only the 6 month issue, but the results would probably be less accurate.

CALCULATION OF SPOT RATES

Let $y_1 = $ the 6-month T-bill rate, and let $y_2 = $ a 1-year Treasury STRIP rate. (Note: Before the Treasury stopped issuing 1-year T-bills, we would have used the 1-year T-bill rate. Remember also that STRIPS's are zero-coupon treasuries.)

Then, for a $1\frac{1}{2}$-year Treasury, we have

$$P = \frac{c}{1+\frac{y_1}{2}} + \frac{c}{\left(1+\frac{y_2}{2}\right)^2} + \frac{c+100}{\left(1+\frac{y_3}{2}\right)^3} \qquad \text{Equation 19.1}$$

where

$$p \qquad = \text{Market price of the Treasury, expressed as a percent}$$
$$C \qquad = \text{Coupon rate of the Treasury, expressed as a percent}$$
$$y_1 \text{ and } y_2 = \text{Known 6-month and 1-year year yields, expressed as a percent}$$
$$y_3 \qquad = \text{Desired yield}$$

Solving, we have

$$y_3 = 2 \left\{ \left[\frac{c+100}{P - \dfrac{c}{1+\dfrac{y_1}{2}} - \dfrac{c}{\left(1+\dfrac{y_2}{2}\right)^2}} \right]^{\frac{1}{3}} - 1 \right\} \qquad \text{Equation 19.2}$$

Generalizing, we have the equation for y_n:

$$y_n = 2 \left\{ \left[\frac{c+100}{P - c \displaystyle\sum_{j=1}^{n-1} \dfrac{1}{\left(1+\dfrac{y_j}{2}\right)^j}} \right]^{\frac{1}{n}} - 1 \right\} \qquad \text{Equation 19.3}$$

Note that in these equations, $j =$ the number of half years from present, not the number of whole years.

You can produce a spot yield curve from this equation. To price a bond, apply the spot rate for each period to its respective bond payment of interest or principal.

USING THE TREASURY SPOT RATE IN THE TREASURY MARKET

You can use the spot rates to evaluate any Treasury security. Simply use the applicable spot rate to evaluate each individual coupon and principal payment. Then compare the price you have computed against the actual market price.

USING SPOT RATES WITH OTHER BONDS

Suppose you own a bond of lesser quality than Treasuries and presumably a higher yield than Treasuries of comparable maturity, and you want to compare the yield on your bond to the yield or yields on Treasuries. Here are two ways you might do it.

Suppose you start with a corporate bond with individual coupon payments c_c and a price P_c. Note that the coupons are paid semiannually, so that the stated coupon rate is $2c_c$.

1. For the first method, you apply an equal amount of spread s over each Treasury spot rate y_i, where the y_i's are the semiannual spot rates you have previously computed. We then have the equation

$$P_c = \frac{c_c}{\left(1+\dfrac{y_1+s}{2}\right)} + \frac{c_c}{\left(1+\dfrac{y_2+s}{2}\right)^2} + \ldots + \frac{100+c_c}{\left(1+\dfrac{y_n+s}{2}\right)^n} \qquad \text{Equation 19.4}$$

Use numerical analysis techniques, such as the bisection method, to solve for s.

2. A second method is to apply some adjustment, or equation, to the spot rates. The adjustment could have many formats, but only one variable is allowed because you only have one equation. For example, you might increase each spot rate by a specified percentage. Let x be the specified percentage. Then the equation would be

$$P_c = \frac{c_c}{\left(1+\dfrac{y_1+x \times y_1}{2}\right)} + \frac{c_c}{\left(1+\dfrac{y_2+x \times y_2}{2}\right)^2} + \ldots + \frac{100+c_c}{\left(1+\dfrac{y_n+x \times y_n}{2}\right)^n}$$

$$\text{Equation 19.5}$$

The individual y entries are $(y_i + [x \times y_i])$ instead of $(y_i + s)$. Use numerical analysis to solve for x.

In Case 1, you are saying, "We are evaluating this bond at s basis points over the Treasury spot rates." In Case 2, you are saying, "We are evaluating this bond at x percent over the Treasury spot rates."

We have just applied this procedure to a bond, but actually you could apply the technique to evaluate any flow of funds, including a project's flow of funds. In Chapter 4, we discussed using a given interest rate to evaluate a flow of funds and we mentioned that we didn't need to use the same rate for each payment. We have now expanded this idea to use a set of rates we developed from the Treasury market. You could then set management requirements for approval of a new project in terms of some relationship to the Treasury spot rate.

IMPLIED FUTURE FORWARD RATES

We can use the expectations hypothesis to derive implied future forward rates.

Let $_nF_t$ = Forward rate n years from now for t years

This means that n years from now, the spot rate for t years will be $_nF_t$.

Suppose an investor wants to make an $(n + t)$ year investment. She can simply buy an $(n + t)$-year security, or she can buy an n year security and then reinvest the proceeds n years from now in a t-year security. The reinvestment would be at the forward rate for t years n years from now.

For the $(n + t)$ year investment, after $(n + t)$ years, the investor will have

$$(1 + y_{n+t})^{n+t}$$

If the investor makes an n-year investment and at year n reinvests for t years at the t year rate n years from now, the investor will have

$$(1 + y_n)^n (1 +_n F_t)^t$$

Using the expectations hypothesis, the investor is indifferent, so we have

$$(1 + y_n)^n (1 +_n F_t)^t = (1 + y_{n+t})^{n+t} \qquad \text{Equation 19.6}$$

Solving for $_nF_t$, we have the implied future forward rate

$$(1 + y_n)^n (1 +_n F_t)^t = (1 + y_{n+t})^{n+t}$$

$$_nF_t = \left[\frac{(1 + y_{n+t})^{n+t}}{(1 + y_n)^n} \right]^{\frac{1}{t}} - 1 \qquad \text{Equation 19.7}$$

Note that in this equation,

y_i are annual spot rates and
$_nF_t$ are annual forward rates

Conversion to semiannual rates is straightforward.

OTHER TERM STRUCTURE HYPOTHESES

We have used the expectations theory of the term structure of interest rates to generate some equations about present and expected future rates. However, other theories of the term structure of interest rates exist. You should know about the main ones, if only to understand that they exist and some of their implications.

The expectations hypothesis is consistent with the normally ascending yield curve if the market expects that interest rates will rise. Interest rates did, in fact, generally rise from the mid 1940s to about 1980. Since then, interest rates have generally declined.

RISK PREMIUM HYPOTHESIS

The risk premium hypothesis states that investors in longer-term securities require a premium for the increased risk in investing in these long-term securities. This premium is stated as an increase in the yield the securities must offer to make them marketable.

The risk premium hypothesis is also consistent with the normal ascending yield curve. The hypothesis is also consistent with the interest in duration. Duration measures at least one kind of risk and increases as time to maturity increases.

LIQUIDITY PREFERENCE HYPOTHESIS

The liquidity preference hypothesis states that longer-term securities are less marketable and less liquid than shorter-term securities. Investors therefore require an additional yield to compensate them for the decreased liquidity. This is similar to the risk premium hypothesis. It is also consistent with the normal ascending yield curve.

MARKET SEGMENTATION HYPOTHESIS

The market segmentation hypothesis says that different types of investors prefer different maturity ranges to answer their particular investment needs. For example, life insurance companies have long-term contracts, with large reserves required for their permanent plan policies (whole life, limited payment plan, and endowment policies), as well as for annuities. However, many, and probably most, investors will move away from their preferred maturity range if the offered yields in other maturity ranges are high enough.

DISCUSSION OF THE VARIOUS HYPOTHESES

All of these hypotheses sound plausible, and all reasonably represent human behavior in at least some parts of the fixed-income investment area. Also, they are not inconsistent with each other; they could all hold simultaneously.

Here are some statements that I have heard during my experience in the fixed-income markets, with a comment on which of the hypotheses it represents in operation:

1. "I'm buying short term because I think money rates are going to rise." This statement is consistent with the expectations hypothesis.
2. "I moved from 2 years to 5 years and picked up 100 basis points in yield." This statement is consistent with the risk premium hypothesis and represents an implicit use of duration as a risk measure.
3. "The Farm (State Farm Insurance Company) is buying." State Farm Insurance Company is a major insurance company, active in the fixed-income market. This statement is consistent with the market segmentation hypothesis.
4. "I'm not being paid enough to move beyond 10 years." This statement is, once more, consistent with the risk premium hypothesis and has an implicit use of duration.
5. "It's too hard and too expensive to sell the long-term bond." This statement is consistent with the liquidity preference hypothesis and also, to some extent, with the risk premium hypothesis.

You can see that all of the hypotheses are represented. Perhaps all of them have some validity. Or perhaps we simply haven't invented (or discovered) a really good theory on the term structure of interest rates.

CHAPTER SUMMARY

The term structure of interest rates is the relationship between the time to a bond's maturity and the yield. Different types of yield curve include the following:

Normal—yields increase as time to maturity increases
Inverted—yields decrease as time increases
Flat—yields stay the same
Hump-backed—yields increase for a while, then decrease

The expectations hypothesis is that the term structure contains within itself the market's expectation of future rates Use this hypothesis to compute spot rates.

The equation for spot rates is as follows:

$$y_n = 2 \left\{ \left[\frac{c+100}{P - c \sum_{j=1}^{n-1} \dfrac{1}{\left(1+\dfrac{y_j}{2}\right)^j}} \right]^{\frac{1}{n}} - 1 \right\}$$

The equation for static yield curve is as follows:

$$P_c = \frac{c_c}{\left(1+\dfrac{y_1+s}{2}\right)} + \frac{c_c}{\left(1+\dfrac{y_2+s}{2}\right)^2} + \ldots + \frac{100+c_c}{\left(1+\dfrac{y_n+s}{2}\right)^n}$$

Another model is this one:

$$P_c = \frac{c_c}{\left(1+\dfrac{y_1+x \times y_1}{2}\right)} + \frac{c_c}{\left(1+\dfrac{y_2+x \times y_2}{2}\right)^2} + \ldots + \frac{100+c_c}{\left(1+\dfrac{y_n+x \times y_n}{2}\right)^n}$$

The equation for implied forward rates is as follows:

$$_n F_t = \left[\frac{(1+y_{n+t})^{n+t}}{(1+y_n)^n} \right]^{\frac{1}{t}} - 1$$

Other term structure hypotheses are the risk premium hypothesis, the liquidity preference hypothesis, and the market segmentation hypothesis.

COMPUTER PROJECTS

1. Use your favorite computer mathematics program to analyze a recent complete Treasury market report, as presented in one of the major national newspapers, to bootstrap a Treasury spot yield curve. How closely does it follow the yields shown in the market report?
2. Take a corporate bond with a higher yield, as quoted in the same newspaper. Use your favorite computer mathematics program to

develop a static yield curve and a yield curve using a percentage increase over the Treasury spot yield curve, as discussed in this chapter. How similar are they?

3. Compare the two. How do they compare with the simple yield to maturity shown in Chapter 6? Is the analysis worth doing in terms of added value? Why?

TOPICS FOR CLASS DISCUSSION

1. Discuss some different equations you might use to develop a set of yields for higher-yielding bonds, based on the Treasury spot yield curve and similar to the static yield curve and the percentage increase yield curve? Which do you prefer and why?

2. You want to develop different yields for the final maturity amount and its associated coupon payment. Why would you want to do this? How would you go about developing these yields?

PROBLEM

Use a complete Treasury market report to compute the 1.5-year spot rate, given the spot rates for 6-month and 1-year zero-coupon Treasuries.

SUGGESTIONS FOR FURTHER STUDY

The following book has an excellent chapter on the term structure of interest rates. It is a well-known and highly regarded book in its field, as shown by the number of editions it has gone through. Van Horne, James C., *Financial Market Rates and Flows*, 5th ed., Prentice-Hall, Englewood Cliffs, NJ, 1997.

CHAPTER 20

Variable and Uncertain Cash Flows

In this chapter, we apply the probability concepts studied in Chapter 18 and the spot rate curve concepts studied in Chapter 19 to learn how to evaluate flows of funds that vary in size, are uncertain, or both. Like the previous chapter, this represents a change in our viewpoint from the earlier chapters. Previously, we assumed that all payments would be made in full, when due, and based all the calculations on that assumption. We now change that to allow for the chance that a payment may not be made. This represents a change, just as in the previous chapter we introduced the idea of differing interest rates for different bond payments. We used varying interest rates, another change from the early part of the book.

We first look at a series of examples, using a simple flow of funds. You should be able to expand this to longer and more complicated fund flows without any real problem.

We apply a single interest rate to evaluate a varying flow of funds. We then apply the probability concepts to this varying flow of funds and compute the value of the flow. We apply varying interest rates to the variable flow of funds and then apply a combination of varying interest rates and probabilities to the same variable flow of funds.

We apply the results to several different business applications, including insurance and project analysis.

Mostly, this chapter, and the next two chapters, don't introduce new concepts but rather apply concepts previously discussed to new situations. When you finish this chapter, you should understand how to apply probability theory and varying interest rates to compute the present value of an uncertain and variable flow of funds.

This analysis can also guide management attention to the flows of funds that might particularly repay this management attention. These might include particularly large fund flows or fund flows with only a chance of being paid. Management might monitor these flows in particular to make efforts to improve the chance of being paid, to increase the size of the payment, or to ensure that the payment will be made.

VALUING A VARYING SERIES OF CASH FLOWS USING THE SAME INTEREST RATE, VARYING INTEREST RATES, AND PROBABILITIES

The first valuation equation is simply the material covered in Chapter 4 on present values. You compute the present value factor for each of the payments and multiply the payment by the factor. Then add the present values of the payments.

The present value of the flow is given by

$$\sum_{j=1}^{n} M_j \times \left(\frac{1}{1+i}\right)^j \qquad \text{Equation 20.1}$$

where

$$M_j = \text{Payment due at time } j$$
$$i \ = \text{Interest rate used}$$
$$j \ = \text{Number of the period}$$

EXAMPLE 20.1.   Here is a varying flow of funds, evaluated at a 6% annual yield:

Year	Payment	Percentage yield	Present value	Present value of payment
1	1	6.0	.9434	0.9434
2	3	6.0	.8900	2.6700
3	9	6.0	.8396	7.5564
4	5	6.0	.7921	3.9605
5	2	6.0	.7473	1.4946
Totals	20			16.6249

Note that almost half the present value of the flow consists of the payment due in year 3. Management might pay particular attention to that particular payment.

To evaluate a flow of funds with probabilities of being paid (or not), multiply each present value by the probability of happening. Here is the equation for its value:

$$\sum_{j=1}^{n} M_j \times p_j \times \left(\frac{1}{1+i}\right)^{j} \qquad \text{Equation 20.2}$$

where

M_j = Payment due at time j
i = Interest rate used
j = Number of the period
p_j = Probability of the payment due at time j actually being paid ($0 \le p_j \le 1$)

∎

EXAMPLE 20.2.   Here is the same varying flow of funds that is shown in Example 20.1, except in this case we are not certain that some payments will actually be made.

Year	Payment	Percentage yield	Present value	Probability of being paid	Present value of payment
1	1	6.0	.9434	1.0	0.9434
2	3	6.0	.8900	0.9	2.4030
3	9	6.0	.8396	0.7	5.2895
4	5	6.0	.7921	0.4	1.5842
5	2	6.0	.7473	1.0	1.4946
Totals	20				11.7147

Note that the present value of the flow has gone down, as expected, because some flows might not be received. The largest payment, 9 in year 3, also has only a .7 chance of being paid, resulting in a large decrease in the present value of that fund flow. However, it still has the largest present value, amounting to almost half the value of the entire flow. Management could easily want to direct its attention to this flow, possibly to increase the probability of being paid.

∎

Here is the same flow, but the payment in year 5 may now be either 1, with a probability of .3, or 2, with a probability of .4. Here is the revised schedule:

EXAMPLE 20.3.

Year	Payment	Percentage yield	Present value	Probability of occurring	Present value of payment
1	1	6.0	.9434	1.0	0.9434
2	3	6.0	.8900	0.9	2.4030
3	9	6.0	.8396	0.7	5.2895
4	5	6.0	.7921	0.4	1.5842
5	1	6.0	.7473	0.3	0.2242
	2	6.0	.7473	0.4	0.5978
Totals	20				11.0421

■

SOURCES OF THE PROBABILITIES YOU USE

Where do you get the probabilities to use? You use the same sources discussed in Chapter 18 or the sources of probabilities used in this kind of work. For example, in doing a project analysis, you might use the probabilities associated with the particular kind of project you are analyzing.

To evaluate a flow of funds with varying interest rates and no payment concerns, use this equation:

$$\sum_{j=1}^{n} M_j \times \left(\frac{1}{1+i_j}\right)^j \qquad \text{Equation 20.3}$$

where

M_j = Payment due at time j
i_j = Interest rate used for funds paid in period j
j = Number of the period

Now we look at the same varying flow of funds as in Example 20.1, but we use different interest rates to compute the present values.

EXAMPLE 20.4.

Year	Payment	Percentage yield	Present value	Present value of payment
1	1	6.0	.9434	0.9434
2	3	6.5	.8817	2.6451
3	9	7.0	.8163	7.3467
4	5	7.5	.7488	3.7440
5	2	8.0	.6806	1.3612
Totals	20			16.0404

■

SOURCES OF THE INTEREST RATES YOU USE

Where do you get the yields to use? You could use the spot yields from the Treasury spot yield curve, and you could adjust them, as shown in the previous chapter, using one of the several techniques discussed. For example, you could add 400 basis points (four percentage points) to the Treasury spot yield curve. You could simply double the spot yield curve. You could use the yields provided by management. You could use the yields customarily used in your field. For example, some fields might customarily use a 6% interest rate. The yields in the previous example were, of course, simply chosen by me for teaching purposes.

Finally, we use both probabilities and varying interest rates to evaluate the flow of funds. The equation for this is as follows:

$$\sum_{j=1}^{n} M_j \times p_j \times \left(\frac{1}{1+i_j}\right)^j \qquad \text{Equation 20.4}$$

where

M_j = Payment due at time j
i_j = Interest rate used for funds paid in period j
j = Number of the period
p_j = Probability of the payment due at time j actually being paid ($0 \le p_j \le 1$)

Here is an example of both probabilities and varying interest rates used to evaluate a varying flow of funds.

EXAMPLE 20.5.

Year	Payment	Percentage yield	Present value	Probability of being paid	Present value of payment
1	1	6.0	.9434	1.0	0.9434
2	3	6.5	.8817	0.9	2.3806
3	9	7.0	.8163	0.7	5.1427
4	5	7.5	.7488	0.4	1.4976
5	2	8.0	.6806	1.0	1.3612
Totals	20				11.3255 ∎

APPLYING PROBABILITY CONCEPTS TO VALUE A VARIABLE OR UNCERTAIN FLOW OF FUNDS

Suppose you have a flow of funds, but you are not sure whether or not you will actually receive an individual payment. For example, you might share in the profits of a business, which will pay you a fixed amount if the business makes money but will pay nothing if the business has a loss.

EXAMPLE 20.6.   *The lakeside cottage, revisited.* Suppose your friend, in exchange for you helping him with the cottage, offers to pay you $200 instead of 2 weeks free use, each year, for 10 years, during each year that his rental income equals or exceeds $4,000 for the year. Should you accept his offer?

Acceptance depends on the value to you of your week in his cottage and your evaluation of your probability of being paid. Perhaps you would rather have the cash than the use of the cottage.

You analyze his expected income from the cottage. You believe that, for the first 3 years, he has only a 60% chance each year of getting $4,000 or more, an 80% chance each year of receiving at least that amount for the next 3 years, and a 100% chance of receiving at least $4,000 each year for the last 4 years. You decide to use an 8% true annual rate to compute the present values of the expected flow of funds. You then decide whether or not to accept.

In this case, the flow of funds is $200 each year, the term period is 10 years, the discount rate is 8% annually, and the probabilities of receiving the payment are .6 for 3 years, .8 for the next 3 years, and 1.0 for the last 4 years. You can then compute the present value and decide whether to accept the offer.

When, and if, you accept the offer, you are accepting an uncertain cash flow. You might not want to do that; it is up to you. Many people don't want an uncertain cash flow, but others will accept one. It is a personal decision with you. However, in making that decision, you at least have the mathematical tools to help you. You also rely on your neighbor's honesty in making any required payments. ■

DIFFERENT SIZE PAYMENTS WITH DIFFERENT PROBABILITIES OF BEING PAID AT THE SAME TIME

In Example 20.3, we showed a case of two different payments, due at the same date, with different probabilities of being paid. This can be extended to several different possible payments, with different probabilities of being paid.

In Example 20.3, we assigned a probability to each payment and computed the expected present value for each payment. Each present value, in turn, was added into the total present value for the income stream.

You can extend this analysis in several ways. You can have more possible payments, each with its own probability, with the sum of all the payment probabilities no bigger than 1. These payments would depend on the actual project. Here are some examples.

EXAMPLE 20.7.   *The machine with a replacement guaranty.* You have just installed a new machine in your production line. The machine has an expected 10-year life, with a manufacturer's guarantee of full replacement if it fails in the first 2 years. You plan to replace the machine at the 10th year, even if it is still working. What are the probabilities for failure, year by year?

For years 1 and 2, the probabilities of failure are zero because the manufacturer will replace it if it fails. You assign an equal probability of failure year by year for the remaining 8 years, with the machine no longer functional after 10 years.

The probability of failure for each year will be .125 because the machine will be replaced after year 10. ■

EXAMPLE 20.8. *The production machine, revisited.* You learn that machines of this sort, if they don't fail within the first 2 years, will run for 10 years without failure. After 10 years, they have an equal chance for failure during each of the next 10 years and will not be operational after 20 years. You decide to keep the machine going until it fails, but only for 20 years at most.

In this case, the failure percentage is .10 for each year from year 11 to year 20. The replacement guaranty for the first 2 years makes the probability of failure zero for the first 2 years, and the experience provides the failure rates for the next 18 years. These failure rates would be zero for years 3 through 10 and .10 for years 11 through 20.

In actual work of this sort, a variety of models, together with failure and repair statistics, exist. Failure probability tables can be constructed from these statistics. You can apply this sort of analysis in many situations, such as the purchase of a new car with a power train warranty. ■

APPLYING THESE CONCEPTS TO LIFE INSURANCE

In the case of insurance, including life insurance, you have probabilities, developed by actuaries, which you can use in computing premiums and reserves for your insurance and for annuities. These probabilities are shown in mortality tables, similar to the American Experience Table we showed in Chapter 18. We use the same equation, Equation 20.2, for the present value of a flow of funds, each of which may or may not be made. In the case of life insurance, only one payment will be made, upon the death of the insured. In this case, we have

$$\sum_{j=1}^{n} M_j \times p_j \times \left(\frac{1}{1+i}\right)^j \qquad \text{Equation 20.5}$$

where

M_j = Insurance payment made at time j
i = Interest rate used
j = Number of the period
p_j = Probability of the insurance payment to be made at time j actually being paid $(0 \leq p_j \leq 1)$; this will be paid only if the insured dies

In actual insurance company work, actuaries have many shortcuts, including use of special functions called commutation functions. Discussion of these functions is beyond the scope of this book.

Here is an example of a life insurance policy.

EXAMPLE 20.9.   *A life insurance policy.* You are saddened to learn that your friend J will die sometime within the next 5 years. But J wants to make a deal with you. At the end of the year in which J dies, you promise to pay $1 to a person named by J (called J's beneficiary). You have a mortality table, which provides you with probabilities now of J's death in each of the next 5 years, and you can earn 8% on your money. How much should you charge J?

You must charge J enough to pay the claim at the end of the year in which it occurs, but you earn on the amount you have at 8% annually. Table 20.1 shows how to compute this amount. The example assumes 8% earnings and mortality rates of .1, .1, .2, .2, and .4 for years 1 to 5, respectively.

You must charge J $0.75633, or 75.633 cents. Table 20.2 shows the actual flow of funds, year by year.

Remember, in the case of an actual insurance company, you must have a large enough population for the statistics to have an actual business application. You cannot do something like this on just one life or just a few lives. For life insurance, you would need data on many lives, possibly many thousands of lives.   ■

TABLE 20.1   Life Insurance Example of Premium Calculation

Year	Value	Present value @ 8%	Probability of event occurring	Present value of payment
1	1	.925926	.1	.09259
2	1	.857339	.1	.08573
3	1	.793832	.2	.15877
4	1	.735030	.2	.14701
5	1	.680583	.4	.27223
Total			1.0	.75633

TABLE 20.2   Life Insurance Example of Flow of Funds

Year	On hand at start of year	Amount earned (at 8%)	Value of Payment made	On hand at end of year
1	.75633	.06051	.1	.71684
2	.71684	.05735	.1	.67419
3	.67419	.05394	.2	.52812
4	.52812	.04225	.2	.37037
5	.37037	.02963	.4	.0

DISCUSSION OF THE LIFE INSURANCE APPLICATION

You have just computed the premium and displayed the flow of funds for a net single premium life insurance policy on J. A net premium does not include any costs for running the insurance company. It only includes the interest earnings and the mortality costs. Insurance companies must add an amount, called the "loading," to cover the costs of insurance company operation. These include labor costs, taxes, sales commissions, and all the other costs of running an insurance company.

In this case, we used a short mortality table, with just five entries, for teaching purposes. As in the case of the American Experience Table, most widely used mortality tables will go from low ages, such as age 0, to advanced ages, such as age 100 or even higher. This makes the calculations possibly more complicated, but the fundamental principles stay the same.

You can also see why the company makes the payment at the end of the year. The premium calculation assumes that it earns income for the entire year.

But most insurance companies take pride in the prompt payment of claims, especially life insurance claims. How would you compute a correct premium for this situation? You are giving an additional benefit, the immediate payment of the claim, so that the premium must be increased somewhat. How would you make this calculation?

You could divide the yearly periods into shorter periods, such as semiannual, quarterly, or even monthly periods. You would assume that the deaths (statistically) would occur evenly throughout the year. This last assumption won't be quite correct, especially for higher ages with higher mortality rates, because mortality rates tend to rise as age increases.

This leads naturally to the development of continuous functions for life insurance, similar to the development of continuous interest rates in Chapter 3. Many life insurance companies use continuous functions, instead of the discrete functions we used earlier, to compute their premiums and their reserves.

Note that the previous example was developed for teaching purposes. In the real world of life insurance, such a policy would require loading for expenses that would increase the premium, perhaps almost to the face amount of $1. Few people would buy such a policy. Remember the discussion in Chapter 18 on the uses and purposes of insurance. Few peoples have needs that the above policy would answer.

Modern actuarial science study starts with economic utility theory and continuous functions, and it assumes a much higher level of mathematics than this book assumes. Managing multibillion dollar insurance companies may require this level of mathematical skill. However, the methods presented in this chapter will be adequate for the needs of most people.

USING DIFFERENT INTEREST RATES

We used the same interest rate for all payments due. However, we could have used different discount rates for each payment, developing them using methods discussed in the previous chapter.

HOW TO COMPUTE AN ANNUAL PREMIUM
FOR THE INSURANCE

We have just computed the single premium for a life insurance policy. But almost all buyers of life insurance policies pay premiums annually, or more frequently. How would you compute the premiums? Historically, other premium payment periods include semiannual, quarterly, monthly, and even weekly periods.

At the start of the contract, the present value of the expected premiums to be received must equal the present value of the expected death benefits. We use Equation 20.2 again, as follows, with P (for premium) replacing M:

$$\sum_{j=0}^{n-1} P_j \times p_j \times \left(\frac{1}{1+i}\right)^j \qquad \text{Equation 20.2}$$

where

P_j = the premium paid at time j, and $j = 0$ (now) to $n - 1$, because the premium is paid at the start of the year.
i = Interest rate used
j = Number of the period
p_j = Probability of the premium payment due at time j actually being paid ($0 \leq p_j \leq 1$); this will be paid as long as the insured is alive, so it is the probability that the insured is still alive at the start of period j

EXAMPLE 20.10.   *Computing the annual premium for J's policy.* Let x be the annual premium due. Premiums are paid at the beginning of the period, so the present value of the premium flow will be the present value of each premium payment, times the probability of being paid.

Then we have

$$x + .9 \times \left(\frac{1}{1.08}\right) \bullet x + .8 \times \left(\frac{1}{1.08}\right)^2 \bullet x + .6 \times \left(\frac{1}{1.08}\right)^3 \bullet x + .4 \times \left(\frac{1}{1.08}\right)^4 \bullet x = .75633$$

Solving, we have

$$x = .2299214$$

You must charge J $.2299, or 22.99 cents, annually, payable at the start of each policy year.

You can also see that we could have used different interest rates to discount each separate premium payment. This would almost certainly have changed the premium. ■

CALCULATING LIFE INSURANCE COMPANY RESERVES

As time passes, the present values of the benefits to be paid and the premiums to be received both change. Usually, the present value of the death benefits increases, and the present value of the premiums to be received decreases. As a result, the present value of what the insurance company must pay exceeds the present value of what it will receive. How does it meet this obligation? It keeps reserves. The value of the reserves it must have on hand equals the present value of the benefits to be paid minus the present value of the premiums to be received. In the example shown in Table 20.3, this is the amount on hand at the end of the year. These are called net reserves.

Table 20.3 shows the annual cash flows for J's policy with annual premiums. The first entry for each year is the amount on hand at the beginning of

TABLE 20.3 Cash Flow and Reserve Analysis

Year 1	Reserve	.0000
	Premium received	.2299
	Interest earned	.0184
	Claims paid	−.1000
	Reserve—year-end	.1483
Year 2	Premium received	.2069
	Interest earned	.0284
	Claims paid	−.1000
	Reserve—year-end	.2837
Year 3	Premium received	.1839
	Interest earned	.0374
	Claims paid	−.2000
	Reserve—year-end	.3050
Year 4	Premium received	.1380
	Interest earned	.0354
	Claims paid	−.2000
	Reserve—year-end	.2784
Year 5	Premium received	.0920
	Interest earned	.0296
	Claims paid	−.4000
	Reserve—year-end	.0000

Totals might not add due to rounding.

the year. This is also the reserve for the start of the year. For year 1, this amount
is zero. The premium received and the interest income, at 8%, are added, and
the value of the claims paid is subtracted, leaving a net amount, the new
reserve, for the end of the year. This continues year by year until at the end
of the last year, the reserve is zero. No more income or claims are expected.
Note that premium income and claims paid are both based on the probability
of occurrence.

EXPLANATION OF YEAR 2 INCOME
AND EXPENSES

We look at year 2 income and expenses and explain the numbers. Explana-
tions for other years will be comparable.

The year starts with the reserve at end of year 1, .1483. To this we add the
premium received. This is the annual premium, .2299, but claims of .1 were
paid in year 1, so we only receive (.9) (.2299), or .2069. This gives us a total
of (.1483) + (.2069) = (.3552) at the start of year 2. We earn 8% interest
on this amount, or (.08) (.3552) = (.0284). We pay claims totaling (.1000)
for a net amount on hand at the end of year 2 of .2837. This is the required
reserve at the end of year 2. (Some results may not quite conform due to
rounding.)

As expected, at the end of year 5, we have no more claims to pay and expect
no more premium income, so the present values of both expected premium
income and future claims are zero and the required reserve is also zero. The
kind of reserve shown in this example is called a net reserve because it has no
adjustments. Other types of reserves exist.

APPLYING THESE TOTALS TO ACTUAL
INSURANCE COMPANY OPERATIONS

Actual insurance company operations include a relatively high expense level
during the first policy year. Expenses include sales commissions and overrid-
ing commissions paid (which might amount to more than half the first year's
premium), administrative expenses to put the policy on the company's books,
and other initial policy expenses. Many companies, especially smaller companies, adjust reserves to allow for these expenses. These adjustments are studied
in actuarial work, but they are beyond the scope of this book. Basically, they
reduce the reserves for the first year, or the first few years, and make up the
shortfall during later years. Sometimes the shortfall is completely made up at
some future time, and sometimes it is never fully made up.

Life insurance companies also need extra reserves in case death claims become larger than assumed. This happened during the great flu epidemic of 1918. Some life insurance companies had concerns about meeting their life insurance claims.

You can see why life insurance companies have large reserves. These reserves are not "spare cash" to be used for various purposes. The insurance company requires them in order to meet its contractual obligations to its policyholders and their beneficiaries.

In these examples we have used the same interest rate to discount all payments. You could use different interest rates for different payments, depending on when they were due. This would complicate the job, but the equations are basically the same. Computer usage should make the job relatively easy and straightforward.

RESERVES FOR OTHER INSURANCE COMPANIES

Insurance companies in other fields also have reserves. These reserves, to the extent they depend on expected premium income and expected claims, would be developed in a similar manner. Most insurance companies have reserves for claims that have actually accrued but that have not been filed with the company. They are called incurred but not reported (IBNR) reserves. Other reserves exist as well, depending on the particular business of the particular insurance company. The Social Security Administration does not have computed reserves, but does have trust fund as a reserve.

COMPUTING THE VALUE OF A PENSION

Many people receive a pension, usually payable monthly. A pension, or life annuity, is a series of payments to an individual, which continues as long as the recipient lives. It differs from the annuity certain studied in Chapter 5 because it contains a life contingency feature. This means that each payment has a probability that it, and all subsequent payments, won't be made. Some pensions offered by insurance companies contain a guaranty that a certain number of payments will be made. For example, a person might buy a life annuity, payable for her or his entire life, with a guaranty that payments will be made for 10 years.

A corporation or government may give a pension to a retired employee. Usually, the employee must have a certain number of years of service and reach a required age in order to retire. Pensions may vest at an earlier age, or with less service. For example, one of the author's previous employers required

10 years of service and attainment of age 55 as an employee to retire. In this case, attainment of retired status entitled the employee to additional retirement benefits as well as the pension. The most important additional benefit was continuation of low-cost medical insurance, but the employer also provided other small retirement benefits, such as continuation of the matching gifts program.

Employees may be entitled to pension benefits, even if they have not reached retirement age, if they have enough years of service and possibly meet other requirements. In these cases, the pension is said to be "vested." The Employees Retirement Income Security Act (ERISA) requires pension vesting after a certain number of years. However, a vested pension may not, and usually does not, entitle the former employee to other retirement benefits, such as possible medical insurance. It may not even entitle the employee to the amount of the pension that a person in retired status would get. Full consideration of pension rights and requirements is beyond the scope of this book, but all employees should know their retirement benefits. Your employer should provide you with information about them periodically.

Social Security retirement benefits are also life annuities, or pensions. The Social Security Administration calculates the retirement benefit according to the requirements of the Social Security laws and regulations. The evaluation of the retirement benefits is similar to the evaluation of private pensions, except that Social Security benefits have annual cost-of-living adjustments. Many public pensions also have cost-of-living adjustments, but most corporate pensions do not. Occasionally some corporations adjust their pensions to reflect some living cost increases.

To evaluate pensions, you use the equation for an uncertain future cash flow. You might use the same interest rate to discount all future payments or you may use different interest rates. To do the evaluation, you will need a mortality table to figure the death and survival probabilities.

Once more, in computing life insurance and annuity benefits, you must use a mortality table that accurately reflects the experience you can expect from your population. In this case, the population consists of present (and prospective) annuitants. For life insurance policyholders and for annuitants this is particularly important, because annuitants live much longer than life insurance policyholders. This phenomenon, called "anti-selection" by the insurance companies, happens because the annuitant (or policyholder) is the best judge of her or his own health. Few people buy annuities if they don't think they are in tip-top health because, for many annuities, once the annuitant buys the annuity, the cost won't be refunded.

EXAMPLE 20.11. Evaluating the value of an annuity. Your friend J has changed her mind about the deal she wants to make with you. Now she wants to receive a life annuity. At the end of each year, if J is still alive, you will pay her $1. Payments cease when J dies. How much should you charge J?

You must charge J enough to pay her benefits, assuming an interest rate and a mortality table. We will use the same interest rate and mortality table we used to compute J's life insurance premium (see Table 20.4).

You can see that this is similar to Table 20.1 for J's life insurance. The payment is the same, if made, and the discount factors are the same. However, Table 20.1 shows the chance of J dying in that year, whereas Table 20.4 shows the chance of J being alive at the end of the year to receive her payment.

How much should you charge J? You should charge her $2.289515 as a net rate.

Table 20.5 shows the flow of funds for J's annuity, year by year.

TABLE 20.4 Analysis of the J's Annuity

Year	Payment	Present value @ 8%	Probability of event occurring	Present value of payment
1	1	.925926	.9	.833333
2	1	.857339	.8	.685871
3	1	.793832	.6	.476299
4	1	.735030	.4	.294012
5	1	.680583	.0	.000000
Total				2.289515

Totals may not add due to rounding.

TABLE 20.5 Analysis of Funds Flow for J's Annuity

Year 1	Premium received	2.289515
	Interest earned	.183161
	Annuity paid	−.900000
	Reserve—year-end	1.572676
Year 2	Interest earned	.125814
	Annuity paid	−.800000
	Reserve—year-end	.898490
Year 3	Interest earned	.071879
	Annuity paid	−.600000
	Reserve—year-end	.370369
Year 4	Interest earned	.029630
	Annuity paid	−.400000
	Reserve—year-end	.000000
Year 5	Interest earned	.000000
	Annuity paid	−.000000
	Reserve—year-end	.000000

Totals might not add due to rounding.

The construction and analysis are similar to the analysis used in Table 20.3. Note that no payments were made in year 5 because J did not live to the end of the year.

Many pensions have a joint and survivor feature. For example, many corporate pensions pay the retiree a certain amount, but if the retiree has a spouse and dies before his or her spouse, the spouse receives a life annuity, usually for a smaller amount. This joint-and-survivor annuity involves analysis of joint life functions and is beyond the scope of this book. Life insurance actuaries can figure the values of joint annuities.   ■

APPLICATIONS TO PROJECT ANALYSIS

Applying these concepts to project analysis is usually straightforward and may even be similar to the insurance applications just discussed. You estimate the inflows and the outflows, including probability factors for each. You choose interest rates in accordance with your management's instructions by adjusting the spot yield curve, as discussed in Chapter 19; by using industry standard interest rates; or by choosing interest rates from some other source.

You also apply the probabilities, based on your analysis of the project and the job requirements. You might use your experience from other, similar jobs.

You could also compute the reserves required to finish the job. You use an equation similar to the calculation of insurance company reserves, computing the present value of the inflows and the outflows, with the required reserves as equaling the present value of the outflows minus the present value of the inflows. You could have several flows of funds, with differing probabilities.

You can also use the same techniques for purely financial flows of funds. The mathematics will be similar.

APPLICATIONS TO BONDS: WEIGHTED AVERAGE DURATION AND EFFECTIVE DURATION

Suppose you have a bond with an embedded option, such as a call feature or a put feature. Clearly, the option might affect the duration in some way. But in what way and for how much? This new duration is called the effective duration. Several different interpretations and solutions exist for this problem. Here are some ideas. Put these ideas together for a more complete understanding of effective duration.

For bonds with no options, the effective duration is the modified duration. Introducing the option, such as a call option, changes the possible flow of payments and therefore may change the duration. Usually it will shorten the flow, so it will also shorten the duration because duration is a measure of the average time to maturity as well as a measure of risk. You can quantify this in several ways. Here are some ideas to consider.

Effective duration measures the average maturity of the cash flows. It could be a simple weighted average of the modified durations of the bond, priced to the call and priced to maturity. It could be a more sophisticated version of this idea, involving the value of the option feature, discussed more completely in Chapter 23. It could be a measure of the price sensitivity of the bond, and it could be related to the slope of the price-yield curve for the bond, as shown in Chapter 17.

As one measure of effective duration, compute the duration to the call feature and the duration to maturity. As the yield to maturity decreases, it will reach a point where the price to maturity and the price to call become the same. This price is called the "crossover price," and the associated yield is called the "crossover yield." Below this yield, the modified duration to call is the effective duration, and above this yield, the modified duration to maturity is the effective duration.

One possible improvement could estimate the probability of the bond being called. The effective duration is the weighted average of the probability of being called and the probability of not being called.

Let P_C = Probability of the bond being called
$P_M = 1 - P_C$ = Probability of the bond going to maturity
D_C = Duration to call date and price
D_M = Duration to maturity

Then we have

$$\text{Weighted average duration} = D_c(P_c) + D_M(1 - P_c)$$

Clearly this can be extended to cover n call features, $C_1, \ldots, C_n$ inclusive. Then we have

$$\text{Weighted average duration} = \sum_i D_{C_i} \bullet P_{C_i} + D_M \bullet P_M$$

where

D_{C_i} = Duration to call feature C_i
D_M = Duration to maturity
P_{C_i} = Probability of bond being called at time of call feature C_i
P_M = Probability of the bond maturing, and
$\sum_i P_{C_i} + P_M = 1$

This procedure requires judgment in assigning the probabilities of the bond being called, and the weighted average duration is therefore subjective to that extent.

Another way of computing the effective duration is to figure the value of the option and deduct it from the price of the bond. The result can be used to compute the effective duration. This will involve the equation for the option's value, discussed in Chapter 23. However, this equation also requires knowledge of the bond's volatility, the risk-free interest rate, and other parameters. These also require some use of judgment.

Like modified duration, effective duration can be related to the slope of the tangent line to the price-yield curve, and it can also be related to the price sensitivity of the price-yield curve. This follows the calculus derivation in Chapter 17. However, this requires that you know or can estimate the shape of the price-yield curve of the bond with the embedded option. This in turn will usually require some judgment.

You can see that in many cases, measuring effective duration almost always requires some subjectivity in choosing various parameters.

THE ADVANTAGES AND DISADVANTAGES OF EFFECTIVE DURATION

Effective duration uses the embedded options and therefore offers some advantages over modified duration. However, these advantages are, to some extent, more apparent than real. Effective duration usually requires some subjectivity in choices of parameters used in its calculation. Even if the choices are not explicitly made, such as in use of a computer options pricing model, the choices are still there and are made by someone. In the case of the computer options pricing model, the computer programmer or systems designer may have made the choices.

Effective duration considers the option provisions and therefore offers a better measure of risk than modified duration for bonds with embedded options, at least to that extent. It also allows comparisons between bonds with different embedded options or with no options.

However, these depend on subjective assumptions. Erroneous judgments can produce effective durations that are quite wrong. Effective duration is also more subject to market variations or change in time to maturity because of the option feature. It can be especially sensitive to changes in market supply, demand, changes in investor sentiment or preferences, and to special features of a particular issue. The general characteristics of options pricing models will affect effective duration.

No universally accepted definition of effective duration exists.

CHAPTER SUMMARY

The equations for evaluating flows of funds are as follows.
Varying payments, fixed rate:

$$\sum_{j=1}^{n} M_j \times \left(\frac{1}{1+i}\right)^j$$

Varying payments, fixed rate, probabilities:

$$\sum_{j=1}^{n} M_j \times p_j \times \left(\frac{1}{1+i}\right)^j$$

Varying payments, varying rates:

$$\sum_{j=1}^{n} M_j \times \left(\frac{1}{1+i_j}\right)^j$$

Varying payments, varying rates, probabilities:

$$\sum_{j=1}^{n} M_j \times p_j \times \left(\frac{1}{1+i_j}\right)^j$$

COMPUTER PROJECTS

1. Use the life insurance illustrations shown in Examples 20.8 and 20.9. Adjust the interest rate and the mortality rates in various ways. What happens to the net single premium and to the annual premiums? Why?
2. Use a mortality table for the entire range of ages, say age 0 to age 100 (or higher). You could use the American Experience Tables, as shown in Chapter 18. You can also obtain these tables from several sources. Try a local insurance company or one of the departments of the U.S. government, such as the Departments of Commerce or Labor. Several reference books also contain mortality tables. Compute premiums for single premium life insurance for several ages. Then compute premiums for annual and more frequent periods. Take this exercise to the limit by computing premiums for continuous compounding. What happens to the premiums? Why?
3. Now do the same things, using varying interest rates. You might start with Treasury spot rates and then adjust them suitably. What happens to the premiums as you adjust these rates? How important are the

interest rate adjustments for near-term payments compared to the interest rate adjustments for payments further away in time? Why?

TOPICS FOR CLASS DISCUSSION

1. How would you go about selecting interest rates to evaluate net premiums and reserves for a life insurance company? What factors would you consider?
2. Discuss the various definitions of effective duration. Which do you prefer and why?

PROBLEMS

1. Compute the present value of the flow of funds in Example 20.6. Would you take the cash or the use of the cottage? Why?
2. Compute some life insurance premiums and compare them with the published rates of some insurance companies. How much do they differ and why?

SUGGESTIONS FOR FURTHER STUDY

The following is an excellent book on duration and convexity, including an exceptional discussion of effective duration, with some mathematics, by a professional portfolio manager. Livingston, Douglas G., C.F.A., *Bond Risk Analysis: A Guide to Duration and Convexity*, Simon & Schuster (NYIF Corp.), New York, 1990. The above section on effective duration draws heavily on this book.

The Society of Actuaries has produced a text and reference book that is a standard for the actuarial profession. It starts with utility theory and continuous functions and generally requires skills beyond those assumed by this book. However, many with higher levels of mathematical skills may find it useful. Bowers, Newton L., Jr., Gerber, Hans U., Hickman, James C., Jones, Donald A., and Nesbitt, Cecil J., *Actuarial Mathematics*, The Society of Actuaries, Itasca, IL, 1986.

CHAPTER 21

Mortgage-Backed Securities

This chapter covers mortgage-backed securities. It starts with the definition of a mortgage, including the various terms used in discussions of mortgage investments, and the mathematics of mortgage calculations, including the equations to calculate mortgage payments, unpaid balances, and interest rates. The chapter then discusses how a mortgage pool is created and briefly describes some mortgage-backed securities. It looks at prepayment schedules and various prepayment models and examines the TBMA prepayment model, along with its equation. It then explains how duration and probability concepts can be applied to investments in mortgages and mortgage-backed securities.

When you finish this chapter, you should understand how mortgage calculations are done, at least some of the common mortgage investment vehicles, and the mathematics of investing in some mortgage investments.

WHAT IS A MORTGAGE?

A mortgage, of the kind we are discussing here, is a pledge of real property to provide security for a loan. Many different kinds of real property have been mortgaged, including railroad rolling stock and equipment and machinery, but mortgages of the sort we are examining in this chapter are pledges of housing. These are one- to four-family houses usually occupied by the owner who has mortgaged the property to buy it. A typical borrower is a home buyer who has bought the house to live in with his or her family.

Along with the mortgage, the owner has signed a note, which actually signifies the loan and states the terms of the loan. Usually, information about the mortgage is filed with a local government, frequently the county clerk. This is often called recording the mortgage.

Early in the 20th century, many homeowners borrowed to buy their homes, but they only paid interest on the mortgage, sometimes quarterly. When the mortgage came due, they simply assumed a new mortgage, or "rolled over" the maturing mortgage. When bad times came, as in the Great Depression of the 1930s, many homeowners could not pay the maturing mortgage and sometimes could not even pay the interest due. As a result, they frequently lost their homes, which they had sometimes lived in for many years. If they had made principal payments over the years, they would have paid off their mortgage and would not have lost their homes. These events helped lead to the development, during the 1930s, of the level-payment self-amortizing mortgage with monthly payments. This is now the standard form of home mortgage in the United States.

HOW A LEVEL-PAYMENT SELF-AMORTIZING MORTGAGE WORKS

The property owner pays interest and principal monthly, and usually real estate taxes and sometimes property insurance as well. Most fixed-rate mortgages have terms to maturity of 30 or 15 years. These make a total of 360 or 180 payments, respectively.

The payments are usually made to a company called a mortgage service company. The mortgage service company usually does not actually own the mortgages. Instead, the mortgage service firm forwards the payment of interest and principal to the actual owner of the mortgages, deducting a fee for its services.

The mortgage service firm places the tax and insurance payments into an escrow fund and pays the taxes and insurance on the property as these become due. These payments will not be discussed in this chapter.

The property owner may prepay all or part of the mortgage at any time. This is the equivalent of a call feature for a bond and means that mortgage payments have a degree of uncertainty not present in many other fixed-income investments. This uncertainty has important consequences for mortgage-backed securities as an investment.

Frequently, the owner pays "points" when he or she receives the mortgage. Points consist of a reduction of the actual principal amount received, although the stated amount of the loan remains unchanged, and determine the size of the monthly payment. This makes points a kind of call premium, paid if the mortgage is paid off ahead of time.

For example, if a borrower of a $200,000 mortgage paid 2 points, the borrower would have 2%, or $4,000, deducted from the mortgage proceeds, the actual cash he or she received. However, the borrower would still owe $200,000, and the mortgage payments would be computed on that amount.

Partial mortgage payments in advance, or prepayments, do not reduce the payment's size, but instead reduce the term to maturity (the number of payments).

Frequently, the mortgages are insured as to payment of principal by one of several programs offered by the United States government or one of its agencies, the Federal Housing Administration (FHA) or the Veterans Administration (VA), as to payment of principal. The Treasury of the United States guarantees payment of interest and principal on securities based on mortgage pools offered by the Government National Mortgage Association (part of HUD).

THE EQUATION FOR LEVEL-PAYMENT, SELF-AMORTIZING MORTGAGES

The series of mortgage payments consists of a limited number of equal payments, equally spaced in time. This series forms an annuity certain. We studied annuities certain in Chapter 5. The mortgage balance will equal the present value of the annuity stream of mortgage payments.

The equation for the mortgage balance will be

$$M = P\left[\frac{1-\left(\dfrac{1}{\left(1+\dfrac{i}{12}\right)^{12n}}\right)}{\dfrac{i}{12}}\right] \qquad \text{Equation 21.1}$$

where

M = Mortgage balance
P = Payment
i = Nominal annual interest rate on the mortgage
n = Number oFf years payments are to be made

Note that the monthly interest rate = $i/12$, and the total number of payments = $12n$. In the rare cases where payments are made other than monthly, replace the 12 in the equation by the actual number of annual payments. In some cases, the borrower makes a payment every 2 weeks of half the monthly payment. This reduces the length of the mortgage because 26 payments are made annually, so more principal is repaid each year, and the principal is reduced and repaid more rapidly.

Here are some examples of level-payment self-amortizing loans.

EXAMPLE 21.1.   You have borrowed $100,000 to buy a house. You have a 30-year loan at 7% annual interest, with no points and the usual monthly payment. Your monthly payment will be $665.30 for interest and principal. Sometimes this is rounded up to a whole dollar amount, with the last payment reduced to compensate for the slight increase.

The same equation could apply to other loans as well. In fact, it could apply to any level-payment, self-amortizing loan, depending on the actual loan agreement. Here is an example of a (real-life) student loan.   ■

EXAMPLE 21.2.   Catherine has a student loan of $23,000. She will repay the loan with monthly payments starting this month. The term is 10 years (120 payments), and the interest rate is 6.75%. Her monthly payment will be $264.10. This was rounded up to $265 monthly. Probably the last payment, in 10 years, will be reduced somewhat to adjust for the earlier overpayments of $.90. This will depend on the actual terms of the loan.   ■

VARIABLE RATE MORTGAGES

Variable rate mortgages have interest rates that may, and almost always do, vary over the life of the mortgage. The interest rate bears some predetermined relationship to a well-known market interest rate, such as the 1-year Treasury rate or the rate on the 30-year Treasury bond. The mortgage rate is said to be "tied" to the market rate. For example, the mortgage rate might be set at 150 basis points higher than the rate on the 30-year Treasury bond, so that if the T-bond yields 5.5%, the mortgage interest rate will be 7%. (Note that these are nominal annual rates.) The mortgage interest rate will change at some set time, such as annually. Sometimes the mortgage will have maximum and minimum interest rates, and sometimes the interest rate change when it is reset also has a maximum. For example, the variable rates might have a minimum of 5%, a maximum of 9%, and a maximum change (up or down) of 1.5% at each annual reset.

The monthly payment is recalculated every time the mortgage rate changes. The calculation of the new monthly payment uses the same equation as that for a fixed-rate mortgage, but the new mortgage balance, the new time to maturity, and the new nominal annual interest rate are the factors used to compute the new monthly payment. Thus, with variable rate mortgages, the monthly payment will change when the interest rate is changed at the reset time. Of course, if the interest rate is not changed, the monthly payment won't be changed either, unless the borrower has made prepayments. Frequently a prepayment will reduce the monthly payment, because the amount owed is smaller than it would otherwise be. If the interest rate increases by a large enough amount, the payment will still increase, even though the amount owed has decreased.

POINTS

When you pay points for a mortgage loan, you increase your effective borrowing rate. How much? That depends on the number of points and the term of the loan. Here is the equation for the new rate, given the points paid:

$$
M\left(\frac{100-D}{100}\right) = P\left[\frac{1-\left(\dfrac{1}{\left(1+\dfrac{i_D}{12}\right)^{12n}}\right)}{\dfrac{i_D}{12}}\right] \qquad \text{Equation 21.2}
$$

where

M = Mortgage balance
P = Payment
n = Number of years payments are to be made
D = Points as a percentage; for example, for 2 points, D = 2
i_D = Annual interest rate after adjustment for points

M, P, and n are as in Equation 21.1, and D is also given, so we solve the equation for i_D using a numerical analysis method. Using points will increase the actual borrowing rate over the rate without points.

MORTGAGE POOLS

Mortgages are usually not retained by the firm who originally makes the mortgage, called the originator of the mortgage. Instead, the originator either sells the mortgage to a permanent investor or, more commonly, packages a collection of mortgages into a pool. The mortgages in a pool will all have the same general characteristics, including interest rate, maturity date, insurance, and possibly even other characteristics, such as location and homeowner's income. Interests, or shares, in these pools are then sold to permanent investors. In most cases, these interests are set up as securities and are traded in the securities markets.

Two types of securities can be set up using these pools: pass-through securities and pay-through securities.

PASS-THROUGH SECURITIES

Pools with pass-through securities issue securities that indicate a percentage share in the pool. The pool simply passes through to the owners of the fund each owner's pro-rate share in all of the payments received. For example, if you own a 2.5% interest in the pool, you will receive 2.5% of each payment received by the pool, including your share of interest, planned principal payments, and principal prepayments. Thus, all the securities issued by the pool are the same and will have the same CUSIP[1] number, and equal principal amounts of securities will receive the same payments.

The best-known such pools are those offered by the Government National Mortgage Association (GNMA, or "Ginnie Mae"), the Federal National Mortgage Association (FNMA, or "Fannie Mae"), and the Federal Home Loan Mortgage Corporation (FHLMC, or "Freddie Mac").

PAY-THROUGH SECURITIES (COLLATERALIZED MORTGAGE OBLIGATIONS (CMOS))

Pools of pay-through securities are set up to offer a wide variety of individual securities. The securities are all secured by the same mortgage pool, but otherwise can have widely differing characteristics, including coupon rate,

---

[1] A CUSIP (Committee on Uniform Securities Procedures) Number is assigned to each security that uniquely identities that security.

TABLE 21.1   Example of CMO Mortgage Pool Tranches

Tranche	Maturity date	Amount
1	5 years	1,000,000
2	10 years	1,500,000
3	15 years	2,500,000
4	20 years	5,000,000

maturity date, sinking fund requirements, and call features. An extensive description of mortgage-backed securities is beyond the scope of this book, but some understanding is required for an understanding of the mathematics involved.

Many issues of mortgage-backed securities are sold with some bonds of the issue scheduled to be repaid before repayment starts on other bonds of the same issue. These different payment classes are called tranches.

For example, suppose that a mortgage pool has issued $10 million (par amount) of bonds that have the maturity schedule shown in Table 21.1.

Each maturity has a sinking fund to retire all the bonds on its final maturity date. Note also that each maturity, also called a tranche, has been assigned a number. This issue has four different securities, each with its own CUSIP (identification) number.

The sinking fund for each maturity can coincide, at least roughly, with the payment schedules for the underlying mortgages. This means that the interest payments will come from the interest paid on the underlying mortgages, and the sinking fund principal payments will come from the planned principal payments on the mortgages. But what about prepayments on the underlying mortgages?

The prepayments introduce an uncertainty about the cash flow into the mortgage pool and therefore into the bond repayments. The trust indenture, or agreement, for the bonds will specify how the prepayments will be used. The trustee for the bonds will make sure that all payments, both the regular payments of interest and principal, and the principal prepayments, will be used as required by the bond indenture.

The prepayments might be used to repay outstanding bonds, starting with Tranche 1, the bonds nearest to maturity and continuing with each tranche in sequence. This means that the additional payments will reduce the shortest maturity bonds, then the next shortest (Tranche 2) when the shortest are paid off, then the third tranche, and finally the fourth tranche.

However, the prepayments could be scheduled in a different way. The bond contract could specify that all the prepayments could go to redeem the longest maturity bonds, the fourth tranche, until these are repaid. Then the prepayments could go to redeem the bonds of Tranche 3, if any are still outstanding, then Tranche 2, and finally Tranche 1, if any of those bonds are still outstanding.

You can see that the two different plans for allocating the prepayments have also allocated the uncertainty of the prepayments. Plan 1 has allocated prepayment uncertainty to all the tranches, so that the prepayment schedule of all tranches is subject to change if any prepayments occur at all. However, Plan 2 allocates prepayment uncertainty mostly to Tranche 4, because that tranche receives the first prepayments. Tranches 1, 2, and 3 might even be repaid according to the planned schedule if the prepayments don't repay Tranche 4 fully before the final maturity date of Tranche 3.

You can easily see that quite complicated repayment plans can be devised for repayment and, in fact, have been devised over the years. For example, Tranche 4 might be split in two, A bonds and B bonds, with prepayments going to the B bonds and no prepayments going to the A bonds. A special group of bonds could be issued that get all the prepayments. A large number of different types of mortgage-backed securities have been designed, with varying degrees of market success. One purpose of this wide variety of securities is to satisfy a wider market than could be satisfied by bonds paying only principal and interest, with a mandatory sinking fund. Within a broad range, the mathematical treatment of these bonds is similar. Note that whenever you buy, or even consider buying, a mortgage-backed security of any sort, you should always refer to the original prospectus, official statement or other official description, or a similar document to get the complete details of the issue you are examining. Some of these securities can become exceedingly complicated. Successful investment requires a good understanding, not only of mortgage-backed securities, but of the particular mortgage-backed securities under consideration. However, some mortgage-backed securities, such as GNMA securities, which are guaranteed by the United States Treasury, have very high quality, but provide higher income than Treasuries. This higher income can more than compensate for the extra work these securities require.

CASH FLOWS FOR MORTGAGES

Before we look at evaluation of mortgage-backed securities, we need to examine more closely the cash flow for a mortgage, for mortgage pools, and the mathematical models used to describe these flows.

Mortgages can receive cash flows in three different ways. The first is the payment of interest on the mortgage. The rate of interest is stated in the original mortgage contract and is part of the usual monthly payment. The second is the planned principal repayment, which is also part of the usual monthly payment. The third source is prepayment of all or part of the unpaid mortgage balance.

Most mortgages allow the borrower to prepay all or part of the unpaid mortgage balance ahead of schedule, almost always without any penalty. People do this for a variety of reasons. One of the most important reasons is refinancing the mortgage to obtain a lower interest rate. Other reasons include the receipt of a bonus, gift, or inheritance; additional savings to repay the mortgage earlier; sale of the house; sale of other investments; and possibly other reasons.

These prepayments introduce an uncertainty into the cash flow of mortgage securities that isn't present in many other fixed-income securities. Several models have been created to measure and predict prepayments and prepayment speed.

PREPAYMENT MODELS

Individual mortgage prepayments cannot usually be predicted, but several mathematical models for prepayments of groups of mortgages have been used during the past 30 years. These include an average life model, the FHA experience model, and the TBMA (previously the PSA) model.

The average life model was used for many years in pricing GNMA and other pass-through securities. These securities were assumed to have an average life of 12 years.

The FHA experience model was used before interest rates rose sharply during the late 1970s. At that time, rising interest rates changed prepayment rates and made the models obsolete.

The most recent and most widely used model was developed by The Bond Market Association (TBMA), previously the Public Securities Association (PSA), and was called the PSA model. When the PSA became the TBMA, the model was renamed the prepayment speed assumptions model, so it is still the PSA model.

The PSA model computes a constant annual prepayment rate (CPR). The rate starts at 0% annually, increases by .2% each month until it reaches 6% annually (at 30 months), and remains constant thereafter. Here is the equation for the PSA model:

$$
\begin{aligned}
CPR &= (t/30) * 6\% \quad \text{if } t \le 30 \\
CPR &= 6\% \text{ if } t \ge 30
\end{aligned}
\qquad \text{Equation 21.3}
$$

where

t = Number of months since the mortgage originated

Prepayment speeds are measured as a percentage of the PSA. For example, if a mortgage pool has a PSA of 200% and is more than 30 months old, the annual prepayment rate is 12% (= 6% • 2.00).

PSA percentages can be used to change the flow of funds of a pool, or issue of CMO securities, under the terms of the original bond contract. For example, in Table 21.1, suppose all prepayment are used to redeem bonds in Tranche 4. However, the contract might say that if the PSA reaches or exceeds 200%, then bonds in Tranche 3 will receive 10% of the prepayments, with only 90% of the prepayments used to redeem bonds in Tranche 4. This means that if, after 30 months, the annual prepayment rate equals or exceeds 12%, then 10% of the prepayments will be used to redeem bonds in Tranche 3 and only 90% will be used to redeem bonds in Tranche 4. This feature allocates some of the prepayment uncertainty to Tranche 3 and away from Tranche 4. However, all of the tranches have some uncertainty, because prepayments cannot be predicted with absolute certainty. Suppose the prepayment rate exceeds 200% of PSA (12% annual rate) and then goes back down again. Do all the prepayments continue to be 90% to Tranche 4 and 10% to Tranche 3? That depends on the original bond contract.

Some investment banking firms or investment advisory firms may develop their own mortgage prepayment models. These are, naturally, beyond the scope of the book. But if you work with one of these firms, you will probably find that the concepts in this chapter (and this book) will help you use these models more effectively and even help in the development of other models.

MORTGAGE-BACKED INVESTMENT MANAGEMENT: APPLICATION OF DURATION AND PROBABILITY CONCEPTS

It might seem that the prepayment problem makes it impossible to apply fixed-income mathematics concepts to mortgage-backed securities. Actually, using the duration and probability concepts developed earlier, we can estimate the cash flows and the duration, and arrive at a market price based on an applicable yield.

To evaluate the present value of a flow of funds, you must first determine the flow. For mortgages, first take the actual contractual cash flow and apply a model to determine the projected cash flow. At present, this will usually be the PSA model, and the cash flow will be determined by applying a percentage of PSA to the contractual cash flow. You now have the projected cash flow.

Then apply a predetermined yield to determine the present value of the projected cash flow. You now have the value of the mortgage or mortgage pool.

You can apply the probability concepts discussed earlier. For example, you can assign probabilities to different PSA rates. Then compute the different cash flows and their present values, then weigh the results according to the probabilities. For example, you might assign probabilities of .25 to a PSA of 100%, .50 to a PSA of 150%, and .25 to a PSA of 200%.

You can also assign probabilities to different yields, as well as to different PSA rates. For example, if 30-year mortgages are yielding about 7%, you might assign probabilities of .25 to a yield of 6.75%, a probability of .50 to a yield of 6.5%, and a probability of .25 to a yield of 6.25%. However, the probabilities for the PSA percentages and for the yields must conform. For example, increasing PSA speeds imply decreasing yields, because a major driving force for refinancing (and repaying) old mortgages is decreasing borrowing rates for the new mortgages.

The best approach might be to assign probabilities for combined PSA speeds and associated yields. For example, in the preceding case, you might assign a probability of .25 to a combination of PSA of 100% and a yield of 6.75%, a probability of .50 to a combination of PSA of 150% and a yield of 6.50%, and a probability of .25 to a combination of PSA of 200% and a yield of 6.25%. You can then weight the results according to the probabilities for a projected present value of the cash flow.

CHAPTER SUMMARY

A mortgage is a pledge of real property to secure a loan. The mortgages discussed in this chapter are pledges of real estate, usually one to four family houses, occupied by the owner.

A level-payment, self-amortizing mortgage is paid by level payments of interest and principal, which gradually pay off the entire mortgage over the life of the mortgage. In the United States, mortgage payments are usually monthly.

The equation for level-payment, self-amortizing mortgages is:

$$M = P \left[\frac{1 - \left(\dfrac{1}{\left(1 + \dfrac{i}{12} \right)^{12n}} \right)}{\dfrac{i}{12}} \right]$$

where:

M = mortgage balance
P = payment
i = nominal annual interest rate on the mortgage
n = number of years of payments

Variable rate mortgages have interest rates that may vary over the life of the mortgage.

The equation for a mortgage with points paid is:

$$M\left(\frac{100-D}{100}\right) = P\left[\frac{1-\left(\frac{1}{\left(1+\dfrac{i_D}{12}\right)^{12n}}\right)}{\dfrac{i_D}{12}}\right]$$

Where

M = mortgage balance
P = payment
n = number of years of payments
D = points in percent
i_D = annual interest rate after adjustment for points

A mortgage pool contains a collection of mortgages, each with the same general characteristics.

Pass-through mortgage-backed securities pass through to each owner of the pool's securities the owner's pro-rata share of the pool's income from all sources.

Pay-through mortgage-backed securities pay each pool security owner according to the pool's original contract.

The equations for the prepayment speed assumptions (PSA) model are:

$$CPR = (t/30) * 6\% \quad \text{if } t \le 30$$
$$CPR = 6\% \quad\quad\quad \text{if } t \ge 30,$$

where

t = number of months since the mortgage originated

Prepayment speeds are measured in percent of PSA. Apply duration and probability concepts to mortgage-backed securities by determining the cash flow, applying probability concepts if desired, and then computing the duration of the cash flow, selecting a yield for the computation.

COMPUTER PROJECTS

1. Using your favorite mathematics computer program, take a typical mortgage, say a 30-year, 6% mortgage. Compute the true interest rate for increasing points, starting with no points. What conclusions, if any, do you draw from this calculation?
2. Consider a typical mortgage, say a 30-year, 6% mortgage. Use your favorite mathematics computer program to compute the flow of funds for varying percentages of PSA, from 0% (no prepayments) to 300% of PSA. Then compute the durations of the resulting flows of funds. What conclusions do you draw, if any?
3. Consider a typical mortgage, say a 30-year, 6% mortgage. Apply different probabilities for different PSA speeds and their associated yields to develop predicted mortgage fund flows. Then compute the associated present values. What conclusions do you draw, if any?

TOPICS FOR CLASS DISCUSSION

1. What factors would you consider in deciding whether to accept points with a mortgage or to have a mortgage at a higher rate without points?
2. A mortgage pool, outstanding for 5 years, has just had 1% of it outstanding principal prepaid in the last month. What does this imply about the interest rate of the mortgages in the pool, compared to current mortgage borrowing rates?

PROBLEMS

1. You have borrowed $250,000 to buy a house. The loan will be for 30 years, payable monthly, at 7%. What is your monthly payment?
2. In Problem 1, you pay 2 points when you take out the mortgage. What is your true yield, expressed either as a true monthly rate or as a nominal annual rate, compounded monthly?
3. A mortgage pool has been outstanding for 4 years. During the last month, 2% of the principal value was prepaid. What PSA speed is this?

CHAPTER 22

Futures Contracts

This chapter covers futures contracts and some of their mathematics. We start with a mention of cash contracts and a discussion of forward contracts, and we present several examples of how these are used. We then move to a discussion of future contracts and explain how they differ and how future contracts differ from forward contracts. We develop the equation for cost of carry and present and explain the equations for the conversion factors for Treasury security futures.

When you finish this chapter, you should understand the important features of cash, forward, and futures contracts, the cost to carry a future, and the conversion factors for Treasury futures.

CASH, FORWARD, AND FUTURES TRADES

When you buy a security in a normal trade, you expect to take delivery and pay for the security in the usual time, typically three business days from the trade date. This is called a cash transaction. For most individual investors, this is their most common investment transaction. Other types of purchase contracts exist, however.

Suppose you heat your house with oil, and in the summer your oil supplier offers you a deal. You agree to buy your oil, for delivery in the winter, at a price agreed upon now, in the summer. When winter arrives, your dealer will deliver the oil and charge you the price you agreed upon in summer. You agreed, in summer, to buy a product, the heating oil, at a future date, in winter, at a certain price, agreed upon in the summer. This type of contract is called a forward contract. Many oil dealers in the Northeastern United States offer just such a contract, and many of their customers agree to the contract. The customers protect themselves against oil price increases and, more generally, against oil price fluctuations. They frequently pay a fee to enter the agreement.

The futures contract offers the next step in trading now for future delivery. The forward contract was specially designed for you, and both you and the dealer expect that you will actually take delivery of the oil and pay for it. A futures contract is a standard contract that can apply to a large number of trades. The contract trades on an organized exchange, with standard contract expiration dates (called settlement or delivery dates), standard products, standard amounts of the product, quoted prices, exchange guarantees of contract fulfillment, margin availability and requirements, and easy and low-cost trading opportunities.

You can see that the forward contract offers you the advantage of a contract especially designed for you and your needs. However, you cannot change it or get out of it without the agreement of the other party. The futures contract offers the advantages of uniformity, ease of trading, and guaranty of performance. However, it may not be entirely suited to your individual needs.

EXAMPLE 22.1.   The Chicago Board of Trade (CBOT® or sometimes CBT), one of the major futures exchanges, has wheat contracts. The size is 5,000 bushels (minimum) and the expiration dates, in early March of 2002, were in March, July, September, and December of 2002. The CBOT also has corn contracts, with size of 5,000 bushels (minimum) and expiration dates, in early March of 2002, were May, July, September, and December of 2002, and March and May of 2003.   ■

EXAMPLE 22.2.   In early March of 2002, the wheat contract on the Chicago Board of Trade traded at 299 cents per bushel of wheat for the future expiring in December of 2002. The contract size of 5,000 bushels meant that the value of the contract was about $15,000.

You can buy a contract; this is called being "long" the contract (if you weren't already short). You can also sell the contract. If you sell a contract you don't already own, you are said to be "short" the contract. You can sell contracts you already own, and if you sell them all, you are said to "close out" your position in the contract. Similarly, if you are short contracts and you buy enough contracts to have a zero balance, you are also said to "close out" your position.

The open interest is the number of one side (long or short) outstanding of a particular contract. For example, if you buy one future contract, you have increased the open interest by one, if the seller went short. If you bought a previously existing future, there was no change in the open interest. Many futures contracts have an open interest well up into the hundreds of thousands of contracts.

Futures contracts exist for a wide variety of commodities, where they were first used. Commodities that have futures contracts include corn, soybeans, wheat, hogs, pork bellies, cocoa, coffee, sugar, cotton, orange juice, gold, silver, copper, crude oil, and heating oil, among others. Although actual delivery of the underlying commodity of a commodity future is possible and is sometimes done, usually future contracts for commodities are closed out on or before the settlement date.

Futures for commodities are generally, and widely, used for hedging and for speculation. Business operations use hedging in a wide variety of ways. Here are some examples, somewhat simplified for teaching purposes.   ■

EXAMPLE 22.3.   You are a wheat farmer in Kansas. You look at the crop you are now raising and can roughly figure the yield. You look at the wheat futures markets and determine that your farm will earn a profit at the present price of the wheat future. You sell some wheat futures and you are now short some contracts. You have a short position.

At harvest time, you take your harvest and sell it at the local grain elevator. At the same time, you close out your futures position. If the price of wheat has risen, you have a loss on the position, since you must buy it back at a higher price than you sold it for. However, you also make more on the sale of the real wheat than you had expected. The gain on the sale offsets the loss on the short. If the price of wheat falls, you receive less cash than expected from the sale of the real wheat, but you make a profit on the short sale. Once more, the gain and the loss should offset each other. This is called a short hedge or a producer's hedge.   ■

EXAMPLE 22.4.   You are a gold mining company in Canada. You see that the price of gold futures will let you operate at a good profit. You sell some gold futures and continue producing your gold. You have used the producer's hedge, in gold futures, to ensure your continued profitable operation.
                                                                              ■

EXAMPLE 22.5.   You are a baker, producing Ma's Bread (just like your mother used to bake). You need wheat to bake your bread, and you buy this wheat from a local wheat dealer. The price of wheat can fluctuate widely, and you would like to have a steady price, at least to some extent. This will allow you to make long-term contracts with some major grocery chains for sale and delivery of large amounts of Ma's Bread.

You buy some wheat futures contracts and establish a long position. You have just stabilized the cost of your wheat. For your operations, you buy the real wheat as needed. If the price of wheat goes up, you pay more for your real wheat, but you also have a gain on the future. If the price of wheat goes down, you have a loss on your future, but you also pay less for the real wheat you use. This is called a long hedge or a consumer's hedge.   ■

EXAMPLE 22.6.   The oil dealer mentioned in the paragraph on forward contracts will almost certainly hedge his position in futures. He will buy a future for heating oil. If the price of oil goes up, he loses money on his contract with you, but makes a profit on his future. If the price of oil goes down, he makes more money on his sale to you, but loses money on his future.

Theoretically, you could yourself hedge in the futures markets for fuel oil, but the contract is for 42,000 gallons, far more than most homes consume in a year.

All these hedges allow the operator (the farmer, the gold miner, the baker, and the oil dealer in these cases) to go about his or her regular business without worrying so much about price fluctuations of a commodity important to the operator. These operators have transferred the risk of price fluctuation to someone else. That someone else could be a producer or consumer who also wishes to hedge, or it could be a speculator who hopes to profit from a market move in his or her favor. Note that we have here, once more, the transfer of risk, just as we transfer risk by using insurance, the design of certain mortgage-backed securities, and certain derivatives.   ■

THE CROSS HEDGE

The cross hedge occurs when the product you use as a hedge does not have price movements that exactly coincide with the product you are hedging for. As a result, you do not have a perfect hedge. You can frequently, however, hedge to some extent. Here are two examples.

EXAMPLE 22.7.   In producing Ma's Bread, you use an especially good form of wheat called Superwheat. It has special features that make Ma's Bread especially good and salable at higher prices than normal bread. Superwheat trades in its own market, which does not have a future contract. You hedge using the standard wheat future. You have probably protected yourself, but only to some extent. If the Superwheat market doesn't move the same way as the standard wheat market, or to the same extent, you have not hedged yourself perfectly. ∎

EXAMPLE 22.8.   In the early 1980s, a very large Wall Street firm had a very large municipal bond department. Municipal bonds trade in a dealer market, and this particular municipal bond department regularly carried a municipal bond inventory worth well up into nine figures. The department hedged its position with Treasury bond futures. Although both municipal and Treasury bond prices respond to interest rate changes, they don't always move together. This was a cross hedge, and it didn't protect the department fully against price fluctuations. Later, the department could have hedged with the municipal future, but this also might not have been a good hedge. The municipal future is created from appraisal prices of recently issued municipal bonds. If the firm's inventory did not reflect these bonds, the municipal future might have been more of a cross hedge than a perfect hedge.   ∎

THE NEED FOR HEDGING MANAGEMENT

Hedging operations must be continually managed; you cannot just forget about them. The larger production, trading, and manufacturing firms may have whole departments devoted just to hedging and trading operations. In some bond departments, the traders may need special managing to make sure they are strictly hedging. ("I'm not speculating, I'm hedging. Yuk! Yuk! Yuk!") All these operations especially need careful management, as shown by occasional large trading losses in a wide variety of firms.

You can also see that well-run hedging operations reduce risk and reduce potential losses. However, they also reduce potential gains from fluctuations in the commodity or security. Giving up these gains and concentrating on running the business that is the true source of profits is a beneficial effect of hedging, just as paying an insurance premium to avoid large loss is usually a worthwhile business or personal activity.

THE FUTURES CONTRACT

The futures contract is designed to make this trading easy. It is a standardized contract for each commodity or set of commodities. Positions have daily mark to market at the end of the day. This means that positions are evaluated using the latest market prices. Margin calls are made if needed.

Margin requirements are set by the exchanges, but individual brokers may ask for more. If the customer's equity in the account declines, the margin must be maintained, as in stock margin accounts. Less volatile contracts may require less margin.

The exchange, or clearing house, guarantees contract performance. The exchange provides a place to trade, a standard contract, a grievance procedure, and, at least implicitly, some guarantee of member performance, capability, and integrity.

The futures markets allow a wide variety of business operations to manage their businesses more effectively. But in the public mind, the speculative function has probably loomed much larger. Public interest in futures speculation has always existed, from an early Griffith film (*A Corner in Wheat*, 1908), and even earlier, to the more recent silver futures transactions by the Hunt family in the late 1970s and early 1980s. Many, probably most, speculative futures ventures end unhappily, as did the silver futures ventures of the Hunt family. Losses can be fast, devastating, and appallingly large. Some contracts have maximum allowable daily changes, so sometimes it is not even possible to close out a losing position. You can only sit and watch your losses grow.

Futures trading is not a realm for the novice or for someone who cannot afford to lose lots of money fast. Even so, just because a futures transaction loses money, you cannot conclude that the entire operation failed. The futures contract may have been part of a hedging operation that was an overall success, even if the futures part lost money. We saw some examples of that circumstance earlier.

SETTLEMENT OF A FUTURES CONTRACT

Many, probably most, futures contracts are closed out by selling the long positions or covering the short positions so that the net balance is zero on the expiration date. However, even for commodities, some positions settle by actual delivery of the product being traded.

Some products cannot be delivered, such as indexes, and these contracts settle according to the rules of the exchange on which they are traded. Frequently, they are settled by a cash payment.

Contracts that are not closed out, or settled by actual delivery, are settled according to the rules of the exchange on which they are traded.

FINANCIAL FUTURES

Early futures were for commodities, but in 1970, the Chicago Board of Trade (CBOT) began trading in Government National Mortgage Association

(GNMA) futures, and in 1975 it began futures trading in long-term Treasury bonds. Other financial future contracts were introduced later. Many of these contracts have succeeded quite well.

Many financial futures differ from commodity futures contracts in one important way. Many financial futures can settle relatively easily by actual delivery of a suitable security. It only requires an electronic transfer of the ownership, the same as most other ownership transactions in the securities likely to have futures contracts. Thus, a wide variety of financial futures contracts may actually settle by delivery of the underlying security. Usually one of several eligible securities may be delivered, and the seller has a choice of security. Even so, most financial futures are closed out. The CBOT has statistics on settlement by product. A look at these statistics shows that Treasury futures are more likely to settle by delivery than commodity futures. The CBOT estimates that less than 1% of Treasury futures are settled by actual delivery.

Futures on indexes and averages cannot usually be settled by delivery. They are settled on expiration day by a cash transfer. For example, the municipal future is based on appraisals of recently issued municipal bonds. The bonds change as new issues are added to the index, and older ones are deleted. This future can only be settled by cash.

In early March of 2002, futures contracts existed for a variety of financial instruments, market indexes, and currencies. These include Treasury bonds and 2-, 5-, and 10-year Treasury notes, the Dow Jones, the S&P, and the NASDAQ stock indexes, a municipal bond index, and Eurodollar deposits, as well as most of the world's major currencies. Table 22.1 shows a comparison of different types of trades.

HEDGING WITH FINANCIAL FUTURES

Bond dealers and portfolio managers use financial futures for hedging in a variety of ways. Bond dealers can hedge large positions (long or short) in their inventories by suitable futures positions. They can do this especially when large new Treasury issues come to market. Portfolio managers can frequently use futures as part of an overall portfolio management strategy.

TABLE 22.1   Comparison of Various Kinds of Trades

Cash trade	Forward trade	Future trade	Financial future
Exchange trade	Individual trade	Exchange trade	Exchange trade
Expected to settle by delivery	Expected to settle by delivery	Not expected to settle by delivery	Not expected to settle by delivery

Prospective borrowers may also hedge by going short a financial future. This sets their borrowing rate and protects against future interest rate fluctuations. If money rates fall, they lose on the future but can borrow at lower cost. If money rates rise, they must pay more to borrow but gain on the future.

COST OF CARRY

A relationship exists between the cash market price of a bond, the price of the future, the coupon rate on the bond, the financing rate, and the time to the contract expiration. This can be called the cost of carry or the equilibrium point between the futures price and the cash price.

The fundamental idea of this approach is that the cost of the purchase of the bond and holding it until the futures expiry date and the cost of buying the bond by buying the future should be equal.

Let

$$P = \text{Cash market price of bond}$$
$$F = \text{Futures price of bond}$$
$$c = \text{Current yield of bond} = C/P$$
$$r = \text{Financing rate (\%)}$$
$$t = \text{Time to futures delivery date (years)}$$

From the settlement of the future contract, we will pay the price of the bond plus accrued interest:

$$\text{Price of bond} \quad = F$$
$$\text{Accrued interest} \quad = ctP \qquad \text{Equation 22.1}$$
$$\text{Total amount paid} = F + ctP$$

From the settlement of the loan to buy the bond, we will have the repayment of the loan, plus the loan interest:

$$\text{Repayment of loan} = P$$
$$\text{Interest on loan} \quad = rtP \qquad \text{Equation 22.2}$$
$$\text{Total amount paid} = P + rtP$$

If the cash and future prices are in equilibrium, then these two amounts paid should be equal, and we have

$$F + ctP = P + rtP$$
$$0 \quad = F + ctP - (P + rtP) \qquad \text{Equation 22.3}$$
$$F \quad = P[1 + t(r - c)]$$

You can see that as t approaches zero, or if the coupon and borrowing rates approach each other, the price of the bond and the price of the future converge.

CONVERSION FACTORS

The CBOT estimates that less than 1% of the outstanding financial futures contracts actually settle by delivery of a security. However, with an open interest
for each of the major contracts well up into six figures, thousands of contracts
will still settle. In late June of 2002, the open interests for the 2-, 5-, and 10-
year Treasury notes and the long Treasury bond were 104,066; 570,789;
777,001; and 451,396 respectively. Even if 1% of the open contracts settle
by actual delivery, almost 20,000 contracts will settle this way. The CBOT
delivery statistics are generally in line with this estimate.

The seller of a financial future has a choice of a number of Treasuries eligible for delivery. The exact number depends on Treasury issuance. The seller
chooses the security he or she will deliver. The buyer must accept the seller's
choice. Table 22.2 shows at list of eligible securities in late July of 2002,
together with their conversion factors and other information.

Which bond should the seller choose to deliver? If the seller could simply
choose to deliver any security on the eligible list, without any price adjustment, he or she would simply choose the security with the lowest dollar price.
Therefore, the CBOT has a conversion factor associated with each security. The
par amount of the delivery is multiplied by the futures price, and in turn by
the conversion factor to determine the actual amount the buyer must pay to
the seller. Accrued interest is added, as is standard with bond transactions.
Naturally, sellers will choose to deliver the security most advantageous to
them.

EXAMPLE 22.9.   You plan to deliver $1 million par value of Treasury bonds
to settle a short contract, with a delivery date in June of 2002. Your futures
price is 101. You plan to deliver the 5 1/2s due 08/15/2028. The conversion
factor is .9346 for June of 2002 delivery, so you will receive $1 million (par
amount) times 1.01 (futures price) times .9346 (conversion factor) = $943,946
plus accrued interest for your bonds.   ■

EXAMPLE 22.10.   Before you notify the CBOT that you plan to deliver,
you change your mind and plan to deliver the 8 1/8s due 08/15/19. The conversion factor for this bond is 1.2245 for June of 2002 delivery. You will receive
$1 million (par value) times 1.01 (futures price) times 1.2245 (conversion
factor) = $1,236,745 plus accrued interest for your bonds.   ■

Each eligible security has a conversion factor for each of the future delivery dates scheduled. As delivery dates expire and new ones begin, the CBOT
computes new conversion factors. The CBOT uses standard equations to
compute the conversion factors for delivery. We show all the equations (for
Treasury securities) for the sake of completeness, and then we examine one of
them for explanation purposes.

TABLE 22.2 List of Eligible Securities

CBOT 2-Year U.S. Treasury Note Futures Contract

This table contains conversion factors for all short-term U.S. Treasury notes eligible for delivery as of July 24, 2002.

	Coupon	Issue date	Maturity date	Cusip number
1. @	$2\frac{1}{4}$	07/31/02	07/31/04	912828AG5
2.	$2\frac{7}{8}$	07/01/02	06/30/04	912828AE0
3.	$3\frac{1}{4}$	05/31/02	05/31/04	912828AD2
4.	$3\frac{3}{8}$	04/30/02	04/30/04	912828AB6
5.	$3\frac{5}{8}$	04/01/02	03/31/04	912828AA8
6.	$5\frac{1}{4}$	05/17/99	05/15/04	9128275F5
7.	$5\frac{7}{8}$	11/15/99	11/15/04	9128275S7
8.	6	08/16/99	08/15/04	9128275M0
9.	$6\frac{3}{4}$	05/15/00	05/15/05	9128276D9
Number of Eligible Issues:	9	5	3	1
Dollar Amount Eligible for Delivery:	$213.0	$119.0	$69.0	$30.0

CBOT 5-Year U.S. Treasury Note Futures Contract

This table contains conversion factors for all medium-term U.S. Treasury notes eligible for delivery as of July 24, 2002.

	Coupon	Issue date	Maturity date	Cusip number
1.	$3\frac{1}{2}$	11/15/01	11/15/06	9128277F3
2.@	$4\frac{3}{8}$	05/15/02	05/15/07	912828AC4
Number of Eligible Issues:	2	2	2	1
Dollar Amount Eligible for Delivery:	$54.0	$54.0	$54.0	$22.0

CBOT 10-Year U.S. Treasury Note Futures Contract

This table contains conversion factors for all long-term U.S. Treasury notes eligible for delivery as of July 24, 2002.

	Coupon	Issue date	Maturity date	Cusip number
1.@	$4\frac{7}{8}$	02/15/02	02/15/12	9128277L0
2.	5	02/15/01	02/15/11	9128276T4
3.	5	08/15/01	08/15/11	9128277B2
4.	$5\frac{1}{2}$	05/17/99	05/15/09	9128275G3
5.	$5\frac{3}{4}$	08/15/00	08/15/10	9128276J6
6.	6	08/16/99	08/15/09	9128275N8
7.	$6\frac{1}{2}$	02/15/00	02/15/10	9128275Z1
Number of Eligible Issues:	7	7	7	6
Dollar Amount Eligible for Delivery:	$138.0	$138.0	$138.0	$126.0

Notes: "@" indicates the most recently auctioned U.S. Treasury security eligible for delivery. The information Chicago Board of Trade as to its accuracy or completeness, nor any trading result, and is intended Trade should be consulted as the authoritative source on all current contract specifications and regulations. Web site at www.cbot.com to subscribe.

Source: Chicago Board of Trade.

(*continues*)

TABLE 22.2 (*continued*)

CBOT U.S. Treasury Bond Futures Contract (See Footnotes, Page 1)

This table contains conversion factors for all U.S. Treasury bonds eligible for delivery as of July 24, 2002. (The next tentatively scheduled update is August 7, 2002.)

	Coupon	Issue date	Maturity date	Cusip number
1.	$5\frac{1}{4}$	11/16/98	11/15/28	912810FF0
2.	$5\frac{1}{4}$	02/16/99	02/15/29	912810FG8
3. @	$5\frac{3}{8}$	02/15/01	02/15/31	912810FP8
4.	$5\frac{1}{2}$	08/17/98	08/15/28	912810FE3
5.	6	02/15/96	02/15/26	912810EW4
6.	$6\frac{1}{8}$	11/17/97	11/15/27	912810FB9
7.	$6\frac{1}{8}$	08/16/99	08/15/29	912810FJ2
8.	$6\frac{1}{4}$	08/16/93	08/15/23	912810EQ7
9.	$6\frac{1}{4}$	02/15/00	05/15/30	912810FM5
10.	$6\frac{3}{8}$	08/15/97	08/15/27	912810FA1
11.	$6\frac{1}{2}$	11/15/96	11/15/26	912810EY0
12.	$6\frac{5}{8}$	02/18/97	02/15/27	912810EZ7
13.	$6\frac{3}{4}$	08/15/96	08/15/26	912810EX2
14.	$6\frac{7}{8}$	08/15/95	08/15/25	912810EV6
15.	$7\frac{1}{8}$	02/16/93	02/15/23	912810EP9
16.	$7\frac{1}{4}$	08/17/92	08/15/22	912810EM6
17.	$7\frac{1}{2}$	08/15/94	11/15/24	912810ES3
18.	$7\frac{5}{8}$	11/15/92	11/15/22	912810EN4
19.	$7\frac{5}{8}$	02/15/95	02/15/25	912810ET1
20.	$7\frac{7}{8}$	02/15/91	02/15/21	912810EH7
21.	8	11/15/91	11/15/21	912810EL8
22.	$8\frac{1}{8}$	08/15/89	08/15/19	912810ED6
23.	$8\frac{1}{8}$	05/15/91	05/15/21	912810EJ3
24.	$8\frac{1}{8}$	08/15/91	08/15/21	912810EK0
25.	$8\frac{1}{2}$	02/15/90	02/15/20	912810EE4
26.	$8\frac{3}{4}$	05/15/90	05/15/20	912810EF1
27	$8\frac{3}{4}$	08/15/90	08/15/20	912810EG9
28.	$8\frac{7}{8}$	08/15/87	08/15/17	912810DZ8
29.	$8\frac{7}{8}$	02/15/89	02/15/19	912810EC8
30.	9	11/22/88	11/15/18	912810EB0
31.	$9\frac{1}{8}$	05/15/88	05/15/18	912810EA2
Number of Eligible Issues:	31	31	30	30
Dollar Amount Eligible for Delivery:	$367.8	$367.8	$356.8	$356.8

The source for all of these equations and the associated examples is the Chicago Board of Trade.[1]

If you look at Table 22.2, you can see that as the coupon increases, the conversion factor also increases. As time from expiration date to maturity decreases, conversion factors below 1 increase, and conversion factors above 1 decrease, like other bond price calculations based on yield. You can also see that conversion factors for 6% securities, at times different from semiannual periods, are 0.9999. This allows for the accrual of interest earnings on accrued interest paid for the security. We discussed this factor in Chapter 6 on bond price calculations.

Conversion factors are calculated using a 6% yield (3% semiannually, the standard convention). This is stated as a level yield curve. This means that the future payments are discounted at 3% semiannually. (The yield previously was 8%, but this was changed to 6% beginning with contracts with March of 2000 expirations.)

CONVERSION FACTOR EQUATION: CBOT U.S. 2-YEAR TREASURY NOTE

The following equation is used to calculate the conversion factors for the 2-year U.S. Treasury note traded at the Chicago Board of Trade. The calculation assumes a 6% yield to maturity.

$$CF = A \times \left[\frac{coupon}{2} + C + D \right] - B \qquad \text{Equation 22.4}$$

where

CF = Conversion factor
coupon = Annual coupon in decimal form
N = Complete years to call or maturity from the first day of the delivery month
z = Number of months in excess of the whole N (rounded down to the nearest quarter for bond and 10-year note futures and to the nearest month for 5-year and 2-year note futures)

[1]The information contained in this publication represents the views and opinions of Robert Zipf and is not necessarily the views of the Chicago Board of Trade (CBOT). The Chicago Board of Trade does not guaranty nor is it responsible for the accuracy or completeness of any information presented in this document.

$$v = z \quad \text{if } z < 7$$
$$\qquad 3 \quad \text{if } z > \text{or} = 7 \text{ (bonds and 10-year note futures)}$$
$$\qquad (z - 6) \text{ if } z > \text{or} = 7 \text{ (5-year and 2-year note futures)}$$

$$A = \frac{1}{1.03^{\frac{v}{6}}}$$

$$B = \frac{coupon}{2} \times \frac{6 - v}{6}$$

$$C = \frac{1}{1.03^{2N}} \quad \text{if } z < 7$$

$$C = \frac{1}{1.03^{2N+1}} \quad \text{otherwise}$$

$$D = \frac{coupon}{0.06} \times (1 - C)$$

EXAMPLE 22.11.   Using this equation, calculate the conversion factor of the 4 1/2s of 01/31/02 U.S. T-note deliverable into the CBOT U.S. 2-year T-note futures contract expiring in March of 2000.

Coupon = 0.0450
N = 1 (March 1, 2000, through March 1, 2001)
z = 10 months (March 1, 2001, through January 31, 2002, rounded down to the next lower whole month)

Then we have

$$v = (z - 6) = 4$$

and we increase $2N$ to $2N + 1$ when we compute C. Computing the factors, including the conversion factor, we have

$$A = \frac{1}{1.03^{\frac{4}{6}}} = .9804870228$$

$$B = \frac{0.045}{2} \times \frac{6 - 4}{6} = .0075$$

$$C = \frac{1}{1.03^{3}} = .915141659$$

$$D = \frac{0.045}{0.06} \times (1 - .915141659) = 0.06364375545$$

$$CF = .98044870228 \times 1.00129 - .0075$$
$$\qquad = .9742474$$
$$\qquad = .9742 \qquad \blacksquare$$

CONVERSION FACTOR EQUATION: CBOT U.S. 5-YEAR TREASURY NOTE

The following equation is used to calculate the conversion factors for the 5-year U.S. Treasury note traded at the Chicago Board of Trade. The calculation assumes a 6% yield to maturity.

$$CF = \frac{1}{1.03^{\frac{X}{6}}}\left[\frac{C}{2}+\frac{C}{0.06}\times\left[1-\frac{1}{1.03^{2N}}\right]+\frac{1}{1.03^{2N}}\right]-\frac{C}{2}\times\frac{(6-X)}{6} \qquad \text{Equation 22.5}$$

where

CF = Conversion factor
C = Annual coupon on \$100 face amount (in decimal form)
N = Complete years to call or maturity
X = Number of months in excess of the whole N (rounded down to full months). For example, if maturity is 4 years, 5 months, then $N = 4$ and $X = 5$. (Note: If $X = 9$, then let $2N = 2N + 1$ and set $X = 3$.)

EXAMPLE 22.12.   Using this equation, calculate the conversion factor of the 4 3/4s of 08/15/04 U.S. T-note deliverable into the Chicago board of Trade's U.S. 5-year T-note futures contract expiring in March of 2000.

C = 0.0475
N = 4 (March 1, 2000, through March 1, 2004, remember, full years)
X = 5 months (March 1 through August 15)

$$CF = \frac{1}{1.03^{\frac{5}{6}}}\left[\frac{0.0475}{2}+\frac{0.0475}{0.06}\times\left[1-\frac{1}{1.03^8}\right]+\frac{1}{1.03^8}\right]-\frac{0.0475}{2}\times\frac{(6-5)}{6}$$

$CF = 0.97567[0.0238+0.7917\times[0.210591]+0.7894092346]-0.00395833$
$CF = 0.9521$

■

CONVERSION FACTOR EQUATION: CBOT U.S.10-YEAR TREASURY NOTE

The following equation is used to calculate the conversion factors for the 10-year U.S. Treasury note traded at the Chicago Board of Trade. The calculation assumes a 6% yield to maturity.

$$CF = \frac{1}{1.03^{\frac{X}{6}}}\left[\frac{C}{2}+\frac{C}{0.06}\times\left[1-\frac{1}{1.03^{2N}}\right]+\frac{1}{1.03^{2N}}\right]-\frac{C}{2}\times\frac{(6-X)}{6} \qquad \text{Equation 22.6}$$

where

CF = Conversion factor
C = Annual coupon on $100 face amount (in decimal form)
N = Complete years to call or maturity
X = Number of months in excess of the whole N (rounded down to complete quarters). For example, if maturity is 8 years, 5 months, then N = 8 and X = 3. (Note: If X = 9, then let 2N = 2N + 1 and set X = 3.)

EXAMPLE 22.13.   Using this equation, calculate the conversion factor of the 4 3/4s of 11/15/08 U.S. T-note deliverable into the Chicago Board of Trade's U.S. 10-year T-note futures contract expiring in March of 2000.

C = 0.0475
N = 8 (March 1, 2000, through March 1, 2008; remember, full years)
X = 6 months (March 1 through November 15), so we set X = 0 and 2N to 2N + 1.

$$CF = \frac{1}{1.03^{\frac{0}{6}}}\left[\frac{0.0475}{2} + \frac{0.0475}{0.06} \times \left[1 - \frac{1}{1.03^{17}}\right] + \frac{1}{1.03^{17}}\right] - \frac{0.0475}{2} \times \frac{(6-0)}{6}$$

$$CF = 1[0.02375 + 0.791667 \times [0.39498355] + 0.60501645] - 0.02375$$

$$CF = 0.9177$$

Source for all these equations and associated examples is The Chicago Board of Trade. ∎

CONVERSION FACTOR EQUATION: CBOT U.S. 30-YEAR TREASURY BOND

The following equation is used to calculate the conversion factors for the 30-year U.S. Treasury bond traded at the Chicago Board of Trade. The calculation assumes a 6% yield to maturity.

$$CF = \frac{1}{1.03^{\frac{X}{6}}}\left[\frac{C}{2} + \frac{C}{0.06} \times \left[1 - \frac{1}{1.03^{2N}}\right] + \frac{1}{1.03^{2N}}\right] - \frac{C}{2} \times \frac{(6-X)}{6} \qquad \text{Equation 22.7}$$

where

CF = Conversion factor
C = Annual coupon on $100 face amount (in decimal form)
N = Complete years to call or maturity
X = Number of months in excess of the whole N (rounded down to complete quarters). For example, if maturity is 22 years, 5 months, then N = 22 and X = 3. (Note: If X = 9, then let 2N = 2N + 1 and set X = 3.)

EXAMPLE 22.14.   Using this equation, calculate the conversion factor of the 5 1/4s of 02/15/29 U.S. T-bond deliverable into the Chicago Board of Trade's U.S. T-bond futures contract expiring in March of 2000.

C = 0.0525

N = 28 (March 1, 2000, through March 1, 2028; remember, full years)

X = 9 months (March 1 through February 15, rounded down from 11 months to the number of months in the nearest full quarter, which in this case is 9 months.

$$CF = \frac{1}{1.03^{\frac{3}{6}}}\left[\frac{0.0525}{2} + \frac{0.0525}{0.06} \times \left[1 - \frac{1}{1.03^{57}}\right] + \frac{1}{1.03^{57}}\right] - \frac{0.0525}{2} \times \frac{(6-3)}{6}$$

$$CF = 0.98533[0.0263 + 0.875 \times [0.814528] + 0.18547193] - 0.013125$$

$$CF = 0.8977$$

∎

UNDERSTANDING THE EQUATIONS FOR COMPUTING CONVERSION FACTORS

Like many fixed-income mathematics equations, these conversion factor equations appear more fearsome than they actually are. We'll look at the one for the 10-year Treasury notes and explain the terms. The other three equations are comparable.

Here is the equation for the 10-year T-note:

$$CF = \frac{1}{1.03^{\frac{X}{6}}}\left[\frac{C}{2} + \frac{C}{0.06} \times \left[1 - \frac{1}{1.03^{2N}}\right] + \frac{1}{1.03^{2N}}\right] - \frac{C}{2} \times \frac{(6-X)}{6}$$

where

CF = Conversion factor

C = Annual coupon on $100 face amount (in decimal form)

N = Complete years to call or maturity

X = Number of months in excess of the whole N (rounded down to complete quarters).

The annual interest rate (yield) is 6%, or 0.06 in decimal form, compounded semiannually. The true yield is therefore 3%, or 0.03, semiannually.

X is rounded down to complete quarters, so X can equal only 3, 6, or 9. If $X = 9$, then set $X = 3$ and increase the 2N factor to 2N + 1. Similarly, if $X = 6$, set $X = 0$ and increase 2N to 2N + 1. You can see that if X is 6 or 9, we increase the number of full periods by 1 (one-half year) and reduce X by 6 (months).

All calculations for this conversion factor are to a quarter. Note that all calculations for 2-year and 5-year notes are rounded down to the next whole month, so all calculations for these conversion factors are to a whole month. These all differ from the standard bond price calculation, which is to an exact day, as shown in Chapter 6, Equation 6.1.

The conversion factor applies to a unit amount (1). The conversion factor is therefore multiplied by the actual par amount of the trade and by the futures contract price, to determine the actual cash for the principal amount of the bonds delivered. Accrued interest is added as well, as usual for bond trades.

The last term inside the outer square brackets $\left(\left[+\dfrac{1}{1.03^{2N}}\right]\right)$ is the present value of the final maturity amount. It shows the future value (1) discounted at 3% for 2N semiannual periods. If the original calculation for X gave $X = 6$ or $X = 9$, then X was reduced by 6 and 2N increased to $2N + 1$.

The middle term inside the outer square brackets is the present value of the annuity of coupon payments, as displayed in Equation 5.6 in Chapter 5 and as shown in the following development:

$$\left[\frac{C}{0.06} \times \left[1 - \frac{1}{1.03^{2N}}\right]\right] =$$
$$\left[\frac{C}{2 \times (0.03)} \times \left[1 - \frac{1}{1.03^{2N}}\right]\right] =$$
$$\left[\frac{C}{2} \times \left[\frac{1 - \dfrac{1}{1.03^{2N}}}{0.03}\right]\right]$$

Once more, if the original value was $X = 6$ or $X = 9$, X was reduced by 6 and 2N increased to $2N + 1$.

We have now the two parts of the standard bond price calculation.

The seller of the bond is entitled to the accrued interest on the bond. The first term inside the outer square bracket, $\left[\dfrac{C}{2}\right]$, is the accrued interest due for one-half year. If the note is priced on an even semiannual period, this is the accrued interest.

The term outside the outer square brackets, $\dfrac{1}{1.03^{\frac{X}{6}}}$, equals 1 if the conversion factor is priced on a semiannual period, because $X = 0$. If the conversion factor is priced to a quarterly period, the factors inside the outer square brackets are discounted back to the previous semiannual period at a 3% semiannual rate.

The last term in the expression, $-\dfrac{C}{2} \times \dfrac{(6-X)}{6}$, adjusts for accrued interest. This adjustment allows for the standard addition of accrued interest to the computed price from the previous term. It corresponds to the third term on the right-hand side of Equation 6.1, the subtraction of accrued interest. If the calculation is to a semiannual period, the accrued interest deduction is 6-months coupon income. If the calculation is to a quarter, the accrued interest deduction is only 3 months coupon income.

You can see that the equations for conversion factors are just a variation of the standard bond price calculation equation that we studied in Chapter 6.

UNDERSTANDING DELIVERABLE GRADES OF TREASURY SECURITIES

Not all Treasuries are eligible for delivery to settle futures transactions. Here are the requirements, as of mid-2002:

> *Long-Term Treasury Bonds.* U.S. Treasury bonds with a maturity of at least 15 years if not callable; if callable, they are not callable for at least 15 years from the first day of the delivery month.
>
> *10-Year Treasury Notes.* U.S. Treasury notes with a maturity at least $6\frac{1}{2}$ years and not greater than 10 years from the first day of the delivery month.
>
> *5-Year Treasury Notes.* U.S. Treasury notes with an original maturity of not more than 5 years and 3 months and a remaining maturity no less than 4 years and 3 months as of the first day of the delivery month. The 5-year Treasury note issued after the last trading day of the contract month is not eligible for delivery into that month's contract.
>
> *2-Year Treasury Notes.* U.S. Treasury notes with an original maturity of not more than 5 years and 3 months and a remaining maturity of not less than 1 year and 9 months from the first day of the delivery month and no greater than 2 years from the last day of the delivery month.

Delivery months are March, June, September, and December.

WEB SITES

The Chicago Board of Trade maintains a Web site at www.cbot.com. This site contains a variety of recent information on various CBOT activities, both current and historical. The CBOT also offers a conversion factor calculation program and a table of conversion factors.

The Chicago Mercantile Exchange maintains a Web site, at www.cme.com. This site contains a variety of information. The Chicago Mercantile Exchange offers a number of interest rate products.

The Board of Trade Clearing Corporation performs a clearing function for a number of exchanges. It maintains a Web site at www.botcc.com. The site offers a variety of information, including historical trading and delivery information.

CHAPTER SUMMARY

Cash trades: Settle by delivery at some short future time.

Forward trades: Made between two individuals, expected to settle by delivery.

Futures trades: Standard contract, executed on an organized exchange. These trades can take long or short positions. Most aren't expected to settle by delivery, and risk is transferred to another party. These trades can be used for hedging or speculation.

Relation between spot and future prices:

$$F = P[1 + t(r - c)]$$

Treasury security futures: Sometimes settle, have conversion factors, and allow delivery of different eligible treasury securities in settlement.

COMPUTER PROJECT

Use your favorite mathematics computer program to enter one of the conversion factor calculation equations. Now adjust the factors for interest rate (yield) and for maturity date. How does the conversion factor change? Why?

TOPIC FOR CLASS DISCUSSION

Why would an exchange, such as the CBOT, choose to round to the next lower month or quarter in the calculation of a conversion factor? What benefits and disadvantages would the exchange and the traders experience with this feature?

PROBLEM

Select an eligible security for delivery and compute the conversion factor. Compare it with the conversion factors from the CBOT. Were they the same? If not, why not?

CHAPTER 23

Options

This chapter discusses some of the various kinds of options and how they are used. It presents the Black-Scholes options pricing model and discusses its assumptions and its key features. It also presents a slightly different mathematical approach for development of the standard compound interest functions. It presents several problems with using Black-Scholes for bond option pricing, and it mentions fractal analysis as an alternative conceptual framework for option evaluation.

The Black-Scholes options pricing model is related to, and can be derived from, the partial differential equation for heat diffusion, first published by the French mathematician Fourier in the early 19th century. Development of this equation, and similar related mathematical work in options, is far beyond the scope of this book and requires mathematical skills and knowledge not assumed by this book. These include a good knowledge of probability, continuous probability distribution functions (especially the normal, or Gaussian, distribution), mathematical statistics, and partial differential equations. However, we can give the reader some idea of what is going on with this equation.

Managers, investment supervisors, and many investors may need only to have a general knowledge of the subject to work successfully in the field. Others will want to learn much more. The field of options is enormous and complicated. Those without much background in mathematics might look at the McMillan book and possibly the Mandelbrot article, listed in the suggestions for further reading at end of this chapter. Others may wish to study the mathematics of options further or to look at one of the mathematical books listed in the reference section.

Studies have shown that most buyers of options lose money. But options can be used for hedging, or insurance, purposes. A hedging, or insurance, operation of some kind may still have an overall success, even if the option part of it loses money. However, probably most speculative buyers of options lose money on their option purchases.

When you finish this chapter, you should have an understanding of how most options work, the main factors in their evaluation, and the assumptions for the Black-Scholes options pricing model. You should also have some understanding of the limits of the Black-Scholes model for bonds and the theoretical problem with Black-Scholes as a description of human behavior in the securities markets.

WHAT IS AN OPTION?

An option, of the sort we are considering here, gives the owner the right, but not the obligation, to make some kind of security transaction. A wide variety of options exist, but we will look only at two of the simple ones, call and put options. Many other options can be constructed by combining various call and put options. Evaluate them by splitting them up into their parts and evaluating the parts separately.

Note the difference between options and futures. Future contracts must be fulfilled in some way, either by closing out the position or actually delivering or receiving the future's underlying commodity or security. Option contracts are fulfilled or not according to the decision of the option's owner.

Most investors probably think of options as applying to stocks and many common stock options are listed on various exchanges. Some bond options are also listed. Options also exist on indexes, including the Dow-Jones and the S&P 500 indexes.

A call option is the right to purchase a security. A put option is the right to sell a security.

The right (to sell or to buy) has a price at which the trade will be done, called the strike price, and a date on which the option will expire, called the expiration or termination date. A very few options have no expiration date, but if they are still outstanding, they were issued years ago, mostly in corporate reorganizations. All actively traded options have an expiration date. The cost of the option is called the premium. In evaluating an option, use is also made of the risk-free interest rate, usually the rate on short-term U.S. Treasury obligations, and the volatility of the security's price.

A difference exists between American and European options. American options allow the option owner to exercise the option at any time up to and including the expiration date. The European option allows the owner to exercise only on the expiration date. An equation exists to convert from the evaluation of one type to the evaluation of the other.

EXAMPLE 23.1.   In mid-April of 2002, an Intel October, 2002, call option with a strike price of 35 traded at $1.30, down $.20 from the previous day's close. Intel itself closed at $28.39 per share that day, down $.55 from the previous day's close.

In this case, the security is Intel common stock, the option is a call, the expiration date is in October of 2002, the strike price is $35 per share, and the premium was $1.30 per share. Intel stock closed at $28.39 the same day, down $.55 from its previous close.   ∎

PURPOSES OF OPTIONS

Like futures, options can be used for hedging and for speculation. Here are some examples.

TABLE 23.1   Option Terms and Definitions

Underlying security	The security the option applies to; sometimes simply called the "underlying."
Strike price	Price at which the option may be exercised.
Premium	Amount paid for the option.
Expiration date	Date on which the option expires, after which it cannot be exercised.
Termination date	Same as expiration date.
American option	Allows the option holder to exercise the option on any date up to and including, the expiration date.
European option	Allows the option holder to exercise the option only on the expiration date.
Risk-free interest rate	A short-term interest rate with no default risk, usually the Treasury bill rate.
Volatility	A measure of how much the security fluctuates.
In the money	The underlying security sells at a price that allows an immediate profit if the option is exercised.
Out of the money	The underlying sells at a price that does not allow an immediate profit if the option is exercised.
At the money	The options sells at a price that gives neither profit nor loss if it is exercised.

EXAMPLE 23.2.   You are long a security, and you think it will go down. You don't wish to sell the security, at least at present. You buy a put, and you have protected yourself against a price decline. If the security goes down, you exercise your put, or simply sell the put, keep the profit on the sale, and keep your security.   ■

EXAMPLE 23.3.   You wish to buy a security and expect the funds sometime soon. You think the security might go up before you get the funds to buy it. You buy a call option. If the security goes up, you exercise your call, or simply sell it and keep the profit.

In both these cases, you have use options as a hedge, or as a kind of insurance, to protect yourself against an adverse market development.   ■

EXAMPLE 23.4.   You own a security. You can develop some extra income by selling a call option on your security; this is called "writing" a call. The premium you receive for the call you sell is additional income to you. However, you give up the chance for a gain if your security goes up in price enough to put the call option in the money, because the call is likely to be exercised and you will have your stock called away from you. In this case, you are using options to generate some extra income. This kind of write, where you already own the security, is called a "covered write," because you own the security.

If you don't own the security, but write the call anyway, you have done a "naked write." Writing naked calls can be dangerous and costly, as the following example shows.   ■

EXAMPLE 23.5.   In late July of 1982, Merrill Lynch stock was trading at about 24.75. The Merrill October 40 call was trading at about $\frac{3}{16}$ (This call gives the owner the right to buy Merrill stock at $40 per share before the expiration date in October, 1980.) This price indicated considerable doubt that the Merrill stock would rise over 15.25 points in less than 3 months. According to rumor at the time, some people were selling the call as a naked write; they did not own the Merrill stock, but were simply selling the call. They expected that Merrill would not rise to 40, the calls would expire without value, and they would pocket $\frac{3}{16}$ (18.75 cents) per call.

Merrill began a strong rise, and by mid-October the stock traded at $47\frac{3}{8}$, almost doubling in price. The call traded at $7\frac{1}{8}$ before it expired, a 38-fold increase from its price just 3 months previous. Anyone who wrote the call as a naked write had to pay out 38 times what he or she received as premium.   ■

Options can also be used for speculation. You simply buy the option, hoping to profit from a market move. Here is an example

EXAMPLE 23.6.   In Example 23.1, you buy the Intel call at $1.30. If Intel sells over $36.30 before the middle of next October, you may make some money.

If the security is trading higher than the strike price of a call, or below the strike price of a put, so that the owner of the option can make an immediate profit by exercising the option, the option is said to be "in the money." If the security is trading lower than the strike price of a call, or above the strike price of a put, so that no profit is made by exercising the option, the option is said to be "out of the money." If the security trades at the strike price, the option is "at the money."   ■

EXAMPLE 23.7.   In Example 23.1, the Intel option is out of the money. If Intel stock rises above 35, the option will be in the money.

You can see that, once more, we have a case of risk transfer. The buyer of a call has transferred the risk of a rise in the security to the writer of the call. (If you are short the security, or simply want to buy it later, and you buy a call, you have protected yourself against a rise in the price of the security.) The buyer of a put has transferred the risk of a fall in the security to the writer of the put. Options exist on many major market averages, so overall market risk can also be transferred.   ■

FACTORS THAT DETERMINE OPTION PRICES

Five factors determine the price of an option:

The security price. As the market price of the security increases, the price of a call option will also increase. The price of a put option will, of course, decrease. For example, if you have an option to call a security at 50 and the price increases from 50 to 60, the value of your option increases.

The option's strike price for the security. As the strike price of the option increases, the price of a call option will decrease, and the price of a put option will increase. For example, if a security is trading at 60, and you own a call option with a strike price of 50, and then the strike price changes to 55, the value of the option has decreased.

The time to option expiration. As the time to option expiration increases, the price of the option will also increase. The option is worth more because you have more time for the security to change in price. For example, a 1-year call on a security with a strike price of 50 is worth more than a 6-month call on the same security, also with a strike price of 50, because there is a greater chance that the security will rise above 50 during the term of the option.

The volatility of the security. As the volatility of the underlying security increases, the price of the option will also increase because a security with greater volatility has a greater chance of putting the option in the money. For example, if Security A is more volatile than Security B and both are trading at the same price, a call or put option on Security A at a given price is worth more than a similar option on Security B at the same price because there is a higher probability that Security A will trade in the money than Security B.

The risk-free interest rate. As the risk-free interest rate increases, the price of a call option will increase. You can hedge the call by owning the security. This is what is done with a covered write. As the risk-free interest rate rises, the interest cost of owning the security will also rise. As a result, the price of the call increases as the interest rate increases. However, usually the call price increase is not great, at least for short-term options.

BLACK-SCHOLES OPTIONS PRICING MODEL

Here is the equation for the Black-Scholes options pricing model.

The theoretical option price $= pN(d_1) - se^{-rt}N(d_2)$ Equation 23.1

with

$$d_1 = \frac{\ln\left(\dfrac{p}{s}\right) + \left(r + \dfrac{v^2}{2}\right)t}{v\sqrt{t}}$$

$$d_2 = d_1 - v\sqrt{t}$$

where

p = Security price
s = Strike price
t = Time remaining until option expiration expressed as a percentage of a
         year
r = Current risk-free interest rate
v = Volatility measured by annual standard deviation
$\ln$ = Natural logarithm
$N(x)$ = Cumulative standardized normal density function

THE ASSUMPTIONS FOR BLACK-SCHOLES

Here are the assumptions for the Black-Scholes equation, as presented in Wilmott, Howison, & Dewynne (1995, page 41), somewhat adjusted for teaching purposes:

1. The asset price follows a lognormal random walk.
2. The risk-free interest rate r and the asset volatility σ are known functions of time over life of option.
3. There are no transaction costs for hedging.

4. The asset pays no dividends.
5. No arbitrage possibilities exist.
6. There is continuous trading in underlying assets.
7. Short selling is permitted, and assets are divisible.

UNDERSTANDING THESE ASSUMPTIONS

Some of these assumptions are assumptions (or statements) about the markets, but two are assumptions (or statements) about human behavior in the securities markets.

Assumptions 3, 4, 5, 6, and 7 are statements about the institutional structure of the securities markets in general or about particular securities markets. How well do the actual markets conform to the assumptions? In some cases, not terribly well, but in other cases, possibly quite well indeed. Markets in many U.S. Treasury securities conform reasonably closely to the assumptions.

Hedging can take place in both the futures markets and in regular markets, at small cost. There are no extra costs for hedging. Assumption 3 is reasonably close.

Assumption 4 is true for European options, but not for American options. A transformation exists to compute the value of the American option from the European option and, alternatively, from the European option to the American option.

Assumption 5 states that the markets are in continuous touch with each other and that no profits, or at least no large profits, can be made by executing trades simultaneously in different markets at the same time. This is generally true of the Treasury markets.

Assumption 6 states that trading goes on at all times. Treasury markets are open worldwide, and trading takes place at all times.

Assumption 7 states that short selling is permitted and that assets are divisible. Short selling can be done in the futures markets and, to some extent, in the regular markets as well. Treasuries are traded in multiples of $1,000, quite small compared to the size of many trades, which are well up into the millions of dollars.

You can see that many Treasury markets conform reasonably well to the Black-Scholes assumptions. Other bond markets, such as the municipal bond market, don't conform closely. For example, short selling in municipal securities, to the extent it is done at all, is difficult and is generally done only in new issues before delivery of the securities. Also, most municipal securities don't exist in large enough quantities for liquid trading markets, such as exist for many Treasuries. Hedging is also difficult or impossible in most municipal bonds.

Assumptions 1 and 2 differ from Assumptions 3 through 7 in one important respect. They are statements about human behavior in the securities markets.

Assumption 2 states that the risk-free interest rate and the volatility of the security are known functions of time over the life of the security. But these can fluctuate as trading takes place in the security. The risk-free interest rate fluctuates daily and can fluctuate widely over a longer period of time, such as a year. The volatility of the security can also change over time, especially as the security is traded more or less frequently, or as public interest in the security increases or decreases. The volatility and the risk-free interest rate could even be random variables. Security price movement theories with random volatility are called stochastic volatility models.

Assumption 1 states that security prices move following a lognormal random walk. We'll discuss later what this means. For now, we will only state that this assumes that price fluctuations in the security are distributed lognormally. This is an assumption about human behavior in the securities markets. It is, in fact, a mathematical model of human behavior in the securities markets. But do people actually behave the way the assumption states? Are price changes, in fact, distributed log-normally? Many at least question this assumption. We'll discuss this assumption further later in the chapter.

If you do work in options, you should at least understand that many observers do not agree with this assumption.

AN IMMEDIATE PROBLEM WITH BLACK-SCHOLES FOR BONDS

In addition to the questions about the trading, the Black-Scholes equation also has another problem as far as bonds are concerned. The model assumes that any positive price has some probability of occurring. But with bonds, the price cannot exceed the total amount of interest and principal to be received over the life of the bond. A higher price would imply a negative interest rate, discussed earlier in the book. For example, a 6% bond with 10 years to maturity could not sell for more than 160, the total amount of interest income and principal it is possible to receive.

This raises questions about the entire validity of Black-Scholes for bonds. At least at the high range of underlying security values, Black-Scholes predicts too high a chance of occurrence. This means that for at least some lower ranges of underlying security values, Black-Scholes predicts too low a chance of occurrence. This could affect the profitability of some investment programs involving options.

HEDGING AND HEDGING RATIOS (THE GREEKS)

Options can be used to hedge portfolios. A variety of hedging techniques and ratios exist. As usual, hedging operations almost always require active management.

THE PUT-CALL PARITY RELATIONSHIP

One way of hedging is called the put-call parity. It works this way.

You have a portfolio Π consisting of one asset S, a long position in a put P and a short position in a call C, with both the put and the call expiring at time T and with the same strike price E. (We follow the notation of Wilmott, Howison, and Dewynne, 1995.) Then

$$\Pi = S + P - C \qquad \text{Equation 23.2}$$

where

Π = Value of the portfolio
S = Value of the asset
P = Value of the put
C = Value of the call

At time T, if Π has fallen in value below E, C is worthless, and the value will be

$$\Pi = S + (E - S) = E \qquad \text{Equation 23.3}$$

If Π has risen in value above E, P is worthless, and the value will be

$$\Pi = S - (S - E) = E \qquad \text{Equation 23.4}$$

In either case, the final value of the portfolio is E. The hedging with a put and a call has set the final value of the portfolio.

If the final value is E, what is the value now? We get that by using the risk-free interest rate r to compute the present value of E at time T, assuming that time now = 0.

$$S + P - C = E e^{-rT} \qquad \text{Equation 23.5}$$

We have set up a relationship between the asset S and its options P and C. This relationship is called the put-call parity relationship. We have eliminated risk entirely. Also, this relationship means that the portfolio does not require any active management. Whether the arrangement earns any income depends on the relative prices and the risk-free interest rate.

HEDGING RATIOS (THE GREEKS)

A number of measures exist to show the relationship between the factors in evaluating the option and the movement of the factors. These are usually represented by Greek letters and are sometimes therefore called "the Greeks."

The delta, or hedge ratio, shows the relative movement between the option and the underlying asset. It measures how much the option will move relative to the underlying. Keeping the hedge ratio steady requires active management. This may also require margin, which will be an additional investment, but will also usually receive interest. The delta is the derivative of the option price relative to the underlying price.

Delta hedging removes most of the risk of random change, but more advanced portfolio management may require additional ratios. These ratios exist, as shown in Table 23.2.

You can see that these ratios will simply be the partial derivative of the option price relative to the variable in the denominator (the second partial derivative for the gamma).

Vega is sometimes called tau, because vega is not a Greek letter. It is sometimes denoted by v, which is actually the Greek letter nu.

Table 23.2 calculates the Greeks by placing the option price in the numerator of the calculation equation. You could compute the Greeks of an entire portfolio Π by placing Π in the numerator instead and taking the partial derivative. For example, you might want to compute the Greeks of the portfolio we created in developing the equation for put-call parity. We would then have the equations shown in Table 23.3.

TABLE 23.2 Hedging Ratios with Symbols and Meanings

Name	Greek symbol	Meaning
Delta	Δ	Option price change as the underlying security price changes. The delta is also sometimes called the hedge ratio.
Gamma	Γ	Delta change as the underlying security price changes.
Theta	Θ	Option price change as time to expiration changes.
Vega	v	Option price change as the underlying security volatility changes.
Rho	ρ	Option price change as the risk-free interest rate changes.

TABLE 23.3 Calculus Expressions for Hedging Ratios for a Portfolio

$\Delta = \delta\Pi/\delta S$	Delta, or hedge ratio, for the portfolio.
$\Gamma = \delta^2\Pi/\delta S^2$	Gamma for the portfolio.
$\Theta = -\delta\Pi/\delta t$	Time decay measure, negative because the original derivative will be negative.
Vega $(v) = \delta\Pi/\delta\sigma$	Sensitivity to volatility.
$\rho = \delta\Pi/\delta r$	Sensitivity to risk-free interest rate change.

ANOTHER MATHEMATICAL APPROACH TO CONTINUOUS FUNCTIONS, AS PART OF THE DEVELOPMENT OF THE BLACK-SCHOLES MODEL

The development of the Black-Scholes model actually starts with the theoretical development of the concept of return. Return is associated with the percentage change of a security price. Therefore,

$$\text{Return} = \frac{dS}{S}$$

The return can be split into two parts, a known, or predictable, return, and a random variance on the return.

The predictable return, also called the drift, $= \mu dt$, with $\mu =$ constant in simple models. In Chapter 3, we developed this concept, assuming that the interest rate was a constant. However, in more complicated models, μ could be a more complicated function, possibly involving many variables.

The second part of the return is a random change, where the

$$\text{Random change} = \sigma dX$$

In this equation,

$$\sigma = \text{the volatility of the security}$$
$$= \text{standard deviation of returns}$$

and dX is a probability, or stochastic, function. Combining these, we have

$$\frac{dS}{S} = \mu dt + \sigma dX$$

This equation generates asset prices because it generates the price changes. It is called a random walk, and it is a stochastic differential equation because it is a differential equation that has a probability function.

If $\sigma = 0$ (the security has no volatility), we have the ordinary differential equation

$$dS/S = \mu dt$$
$$dS/dt = \mu S$$

If μ is constant, $S = S_0 e^{\mu(t-t_0)}$
If $S_0 = 1$, $t_0 = 0$, $t = 1$,

then

$$S = e^{\mu}$$

which we developed in Chapter 3.

Also, dX contains the randomness associated with the asset prices, and is called a Wiener process.

Assume that dX is a random variable, drawn from the standardized normal distribution. Then

$$\text{Mean } dX = 0$$

and

$$\text{Variance of } dX = dt$$

This can be written as

$$dX = \phi\sqrt{dt}$$

where ϕ is a random variable drawn from standardized normal distribution, with mean = 0 and variance = 1.

The solution is

$$\phi = \frac{1}{\sqrt{2\Pi}}e^{-\frac{1}{2}\phi^2}, -\infty < \phi < +\infty$$

This is a lognormal random walk, and it is what mathematicians mean by that term. Note that ϕ applies to the changes in the price of the underlying security, not to the prices of the security itself.

We have expanded the concept of return beyond the simple constant i used in chapters 2–7 in two ways: (1) we have allowed for a more complicated expression for return than a constant, and (2) we have allowed for uncertainty in this return.

OTHER APPROACHES: FRACTAL ANALYSIS

Fractal analysis offers a different model of security price movements. A mathematical development of fractals involves fractional dimensions and other mathematical developments far beyond the scope of this book. However, we can give a brief description of how fractal analysis operates and what it implies about human behavior in the securities markets.

Fractals offer a different method of generating security prices and security price changes. In this respect, they act like probability-generating functions, such as the normal distribution. In the preceding section, we saw how the normal, or Gaussian, distribution generates a log-normal random walk for security price changes.

The security price movements generated by fractals are quite similar to actual market price movements. Even with elaborate analysis, you can hardly tell them apart. This means that fractal analysis may model the market just as well as the normal distribution model, but it also implies that human behavior in the securities markets does not result in lognormal distribution of security price changes. This in turn implies that the Black-Scholes model may not accurately predict security price movements.

In particular, many competent observers believe that the Black-Scholes model understates the probabilities of extreme changes, while fractal modeling may result in better predictions of such changes. Many of these persons believe that a major reason for the problems of the Long Term Capital Corporation in 1999 was the result of using models, including the Black-Scholes model, which did not adequately predict the extreme price movements actually experienced.

If you manage investment activities of any sort, you should be aware of the existence of fractal analysis and what it can mean to your investment activities. If you do mathematical analytical work in securities, you might want to look at one, or all, of the references on fractals at the end of this chapter.

CHAPTER SUMMARY

An option is the right to make some kind of security transaction. The owner of the option decides whether or not to fulfill the transaction.

A call is the right to buy a security.

A put is the right to sell a security.

Strike price is the exercise price of the option.

The expiration (or termination) date is the last day on which the option may be exercised.

The premium is the price paid for the option.

With the American option, the option may be exercised on any date up to and including the expiration date.

With the European option, the option may be exercised only on the expiration date.

Options may be used for hedging, income production, and speculation.

Factors that determine the option price are the security price, the strike price, the time to expiration, the security volatility, and the risk-free interest rate.

The Black-Scholes equation is the standard option pricing (evaluation) equation.

Some problems arise with Black-Scholes: the security volatility may change, and the risk-free interest rate may change.

Black-Scholes requires assumptions about the institutional structure of the market and human behavior in the securities markets.

Black-Scholes has a special problem with bond options because it assumes a probability of bond prices exceeding the total amount of payments due (negative interest rate).

Hedging techniques include the put-call parity relation and the use of hedging ratios. The delta is an important hedging ratio.

Fractal analysis offers an alternative option analysis model with different human behavior assumptions.

COMPUTER PROJECTS

1. Using your favorite mathematics computer program, enter the Black-Scholes options pricing model. Enter the parameters for a recently traded, active option to solve for the price. Then vary the five parameters that affect the price. How much does the options price vary as each parameter is varied? How would this affect your options trading system, if you had one?

2. Using your favorite mathematics computer program, enter the Black-Scholes options pricing model. How would you adjust it so that the probability of a price higher than the total of interest and principal to be received is zero? You could lower the variance of the underlying normal distribution so that the tail over the maximum total amount is acceptably low. What would this do to the overall analysis? You could simply lop off the tail of the normal distribution at the maximum total. Where would you put the amount lopped off? What would this do to the overall analysis? What other adjustments could you make to the equation?

TOPICS FOR CLASS DISCUSSION

1. Your boss has asked you to develop an alternative options pricing model to the Black-Scholes model. How would you go about doing this? Keep in mind that any options pricing model must reflect human behavior in the securities markets.

2. In the development of the compound interest functions, what other functions could you use in the differential equation for future value and return?

3. How would you change the Black-Scholes model to allow for the maximum price of a bond, the total of interest and principal to be received under the contract? What different methods might you use, and what would be the results?

SUGGESTIONS FOR FURTHER STUDY

The following is a popular text on options. Although the book is not mathematical, it covers the subject. Its long-term popularity, shown by four editions, testifies to its value. McMillan, Lawrence, *Options as a Strategic Investment*, 4th ed., Prentice-Hall, New York, 2001.

Many fine books exist that cover options from a mathematical viewpoint. The following are all well-regarded books on options, with a mathematical approach. You should have at least some understanding of calculus and the ability to follow differential equation manipulations to study any of them. I believe that the books by Chriss and by Willmott *et al.* go well together. Hull has also written a very well regarded book.

Chriss, Neil A., *Black-Scholes and Beyond Option Pricing Models*, McGraw-Hill, New York, 1997. Written by an active Wall Street practitioner and NYU Courant Institute adjunct with a Ph.D. in mathematics.

Hull, John C., *Options, Futures, and Other Derivatives*, 4th ed., Prentice-Hall, Englewood Cliffs, NJ, 1997. Meant for graduate or upper-undergraduate students. Widely used as a textbook.

Willmott, Paul; Howison, Sam; and Dewynne, Jeff, *The Mathematics of Financial Derivatives*, Cambridge University Press, New York, 1995. For graduate or advanced undergraduate students. Presents modeling of mathematical derivatives from an applied mathematician's point of view. Used as a textbook in several well-known universities.

The following contains a series of articles on fractals and scaling, written by the leader in the field. Mandelbrot, Dr. Benoit B., Ed., *Fractals and Scaling in Finance*, Springer Verlag, Berlin 1997.

For a quick discussion of fractals in finance, this article is very helpful. Mandelbrot, Dr. Benoit B., "A Multifractal Walk down Wall Street," *Scientific American*, February 1999, pages 70–73.

A standard reference for stochastic differential equations is: Øksendal, Bernt, *Stochastic Differentiations, 4th Edition*, Springer-Verlag, Berlin, 1995.

INDEX